Handbook of Radiobiology

Handbook of Radiobiology

(For Postgraduate Students of Medical Physics, Radiology and Imaging, and Radiation Oncology)

Thayalan Kuppusamy

MSc DipRp MPhil PhD FUICC FIMSA FUSI

Consultant Medical Physicist
Dr Kamakshi Memorial Hospital
Chennai, Tamil Nadu, India

Professor (Retired)
Madras Medical College
Chennai, Tamil Nadu, India

Foreword
PG Gopalakrishna Kurup

JAYPEE The Health Sciences Publisher
New Delhi | London | Philadelphia | Panama

 **Jaypee Brothers Medical Publishers (P) Ltd**

Headquarters

Jaypee Brothers Medical Publishers (P) Ltd
4838/24, Ansari Road, Daryaganj
New Delhi 110 002, India
Phone: +91-11-43574357
Fax: +91-11-43574314
Email: jaypee@jaypeebrothers.com

Overseas Offices

J.P. Medical Ltd
83, Victoria Street, London
SW1H 0HW (UK)
Phone: +44 20 3170 8910
Fax: +44(0)20 3008 6180
Email: info@jpmedpub.com

JP Medical Inc
325 Chestnut Street
Suite 412, Philadelphia PA 19106, USA
Phone: +1 215-713-718
Email: support@jpmedus.com

Jaypee-Highlights Medical Publishers Inc
City of Knowledge, Bld. 235, 2nd Floor, Clayton
Panama City, Panama
Phone: +1 507-301-0496
Fax: +1 507-301-0499
Email: cservice@jphmedical.com

Jaypee Brothers Medical Publishers (P) Ltd
17/1-B Babar Road, Block-B, Shaymali
Mohammadpur, Dhaka-1207
Bangladesh
Mobile: +08801912003485
Email: jaypeedhaka@gmail.com

Jaypee Brothers Medical Publishers (P) Ltd
Bhotahity, Kathmandu
Nepal
Phone: +977-9741283608
Email: kathmandu@jaypeebrothers.com

Website: www.jaypeebrothers.com
Website: www.jaypeedigital.com

© 2017, Jaypee Brothers Medical Publishers

The views and opinions expressed in this book are solely those of the original contributor(s)/author(s) and do not necessarily represent those of editor(s) of the book.

All rights reserved. No part of this publication may be reproduced, stored or transmitted in any form or by any means, electronic, mechanical, photocopying, recording or otherwise, without the prior permission in writing of the publishers.

All brand names and product names used in this book are trade names, service marks, trademarks or registered trademarks of their respective owners. The publisher is not associated with any product or vendor mentioned in this book.

Medical knowledge and practice change constantly. This book is designed to provide accurate, authoritative information about the subject matter in question. However, readers are advised to check the most current information available on procedures included and check information from the manufacturer of each product to be administered, to verify the recommended dose, formula, method and duration of administration, adverse effects and contraindications. It is the responsibility of the practitioner to take all appropriate safety precautions. Neither the publisher nor the author(s)/editor(s) assume any liability for any injury and/or damage to persons or property arising from or related to use of material in this book.

This book is sold on the understanding that the publisher is not engaged in providing professional medical services. If such advice or services are required, the services of a competent medical professional should be sought.

Every effort has been made where necessary to contact holders of copyright to obtain permission to reproduce copyright material. If any have been inadvertently overlooked, the publisher will be pleased to make the necessary arrangements at the first opportunity.

Inquiries for bulk sales may be solicited at: jaypee@jaypeebrothers.com

Handbook of Radiobiology

First Edition: **2017,** Reprint: 2024

ISBN 978-93-86107-43-5

Printed in India

Foreword

It gives me great pride and pleasure in introducing to the concerned readership, the *Handbook of Radiobiology* written by Dr Thayalan Kuppusamy, Professor (Retired) of Madras Medical College, Chennai, Tamil Nadu, India. Though Dr K Thayalan has published a few books related to radiation and its applications in medicine, this book gives very valuable information on radiobiology for professionals, students preparing for examination and even for routine medical practice.

In this book, topics related to interaction of radiation with cells are dealt with, in which beneficial damage to cancer cells due to radiation therapy and hazards leading to injuries to normal cells and organs, are described in detail with illustrations and workbook examples. Topics, such as biological effects and normal tissue reaction of radiation, biological basis of external beam radiotherapy and brachytherapy, time-dose-fractionation methods, and radiation safety and protection, are well explained in simple terms for all sectors of people to understand.

I am sure, this book will be highly beneficial to postgraduate students preparing for MD and DNB examination in radiotherapy and those for MSc Medical Physics. Further, it would be very useful for professionals in their routine clinical practice in radiotherapy for calculation of effective dose due to various treatment protocols followed today.

Dr Thayalan really deserves compliments for compiling such a worthy book on radiobiology in a simple language for all sectors of personnel, both students and staff, who get benefited.

PG Gopalakrishna Kurup MSc DipRP PhD
Chief Medical Physicist
Apollo Specialty Hospital
Chennai, Tamil Nadu, India

Preface

Radiation has been used in medicine, research and development, and industry since the discovery of X-rays in 1895 by the German physicist WC Roentgen. Over the period, its medical application has grown enormously, and it is widely used in radiology and imaging, and in radiation oncology. Radiologists, radiation oncologists and medical physicists have become professionals of radiation in their respective field. Academic institutions are offering courses to provide such human resources which are highly in demand.

The outcome of any diagnostic and therapeutic procedure depends on the radiation interaction with tissue or cells, or simply based on radiobiological principles. Understanding of the above principle is very much essential not only to predict outcome but also to use effectively the given radiation technique or procedure. Hence, an attempt is made to explain the basics of human body, radiation physics, its interactions with tissue and cells, survival curves and its modifying agents, biological effects, biological basis of radiotherapy, and including radiation safety and protection, in the form of a single book. Attempt is also made to incorporate worked examples wherever needed. Large numbers of figures and tables are incorporated, wherever necessary, for better understanding of the concept.

This book is intended for the postgraduate students of medical physics, radiation oncology, and the radiology and imaging. This is a first book of its kind. I am very proud and happy to come out with this book, incorporating my three decades of teaching experience in radiological physics. Constructive comments are invited from the readers for the betterment of the book in its next edition.

Thayalan Kuppusamy

Acknowledgments

I am thankful to my wife Tamilselvi, son Parthiban and daughter Kayal Vizhi, for their support and cooperation during the book writing process. I also thank Dr RK Jeevanram, Scientific Officer (Retired), Indira Gandhi Centre for Atomic Research, Kalpakkam, Tamil Nadu, for having corrected the entire manuscript and offered suggestions. I also thank Dr Kavitha Arunachalam, Assistant Professor, Indian Institute of Technology, for her valuable input. I acknowledge Dr Kamakshi Memorial Hospital management and my Medical Physics colleagues, especially Dr A Murugan, Ms X Sidonia Valas, Ms S Purnima and Mr T Godwin Paul Das, for their untired support and assistance during the book writing process.

I thank Dr PG Gopalakrishna Kurup, Apollo Specialty Hospital, Chennai, Tamil Nadu, India, for having written the foreword to this book. I am also thankful to Shri Jitendar P Vij (Group Chairman), Mr Ankit Vij (Group President), Mr Tarun Duneja (Director-Publishing) and Bengaluru Production Unit of M/s Jaypee Brothers Medical Publishers (P) Ltd, New Delhi, India, for publishing this book as usual in a neat and elegant manner.

Contents

1. **Human Biology**   1
   Radiobiology 1
   Human Anatomy 1
   Cell 4
   Human Body Composition 9
   Molecular Forces and Bonds 15
   Tumor Biology 16

2. **Radiation Physics**   21
   Atomic Structure 21
   Electromagnetic Radiation 24
   Radiation Units 26
   Radiation Interaction with Tissue 27
   Particle Interactions 34

3. **Interaction of Radiation with Cell**   38
   Sequence of Radiation Events 38
   Direct and Indirect Action 39
   Radiolysis of Water 40
   Characteristics of Actions of Radiation 42
   Irradiation of Macromolecules 43
   Effects of Radiation on Cell Division and Growth 47
   Synchronous Cell Cultures 49
   Radiation and Chromosome Damage 50
   Biodosimetry 52

4. **Cell Survival Curves**   54
   Radiation and Cell Death 54
   Survival Fraction and Plating Efficiency 54
   Survival Curve 55
   Repair of Radiation Damage 61
   Mechanism of Cell Death 63
   Classification of Radiation Damage 65

5. **Modification of Cell Survival**   68
   Modifying Agents 68
   Oxygen Effect 68
   Relative Biological Effectiveness 72

Linear Energy Transfer   73
Cell Cycle Stage   76
Radiation Protectors   77
Radiation Sensitizers   79
Hyperthermia   80

6. Biological Effects of Radiation — 84
Radiation Effects   84
Radiation Carcinogenesis   86
Heritable Effects   94

7. Normal Tissue Reactions of Radiation — 99
Deterministic Effects   99
Radiation Effects on Tissues: Early and Late   99
Skin   101
Hematopoietic System   102
Gastrointestinal Tract   104
Reproductive Organs   105
Eye Lens   106
Other Tissues   107
Acute Radiation Syndrome   108
Radiation Effects on Embryo and Fetus   114

8. Biological Basis of External Beam Radiotherapy — 120
Evaluation of Radiotherapy   120
Effect of Dose Fractionation on Cell Survival   121
Effect of Dose Rate on Cell Survival   122
Rationale for Fractionation in Radiotherapy   123
Dose-response Curves   127
Fractionation Strategies   130
Dose Fractionation Effects in External Beam Treatments   133

9. Biological Basis of Brachytherapy — 135
Brachytherapy   135
LDR and HDR Brachytherapy   136
Dose Rate Effect in Brachytherapy   137
Role of 4 Rs in Brachytherapy   139
Clinical Implications   140
Therapeutic Ratio in Brachytherapy   144
Practical Issues in Radiotherapy   145

10. Time, Dose and Fractionation Models — 149
Isoeffect Models   149
Cube Root Rule   149
Nominal Standard Dose   150
Cumulative Radiation Effect   151

Time-dose-fractionation Model 152
Linear Quadratic Model 153
Influence of Radiobiological Parameters 161
Dose-fractionation Effects in Gynecological Brachytherapy 162
Gaps in Radiation Treatment 164

11. Radiation Safety and Protection   **168**
Equivalent Dose and Effective Dose 168
Sources of Radiation 170
Principles of Radiation Protection 172
Regulatory Requirements 175
Personnel Monitoring 178

Index   ***183***

Human Biology

CHAPTER 1

RADIOBIOLOGY

Ionizing radiations such as X-rays and gamma (γ) rays are used in medicine for both diagnostic and therapeutic applications. This is the outcome of discovery of X-rays by Wilhelm Conrad Röntgen (1895), radioactivity by Antoine Henri Becquerel (1896), and radium by Pierre and Marie Curie (1898). Though these radiations are beneficial, they are also associated with harmful effects. The first radiobiological effect was reported by Henri Becquerel, who unknowingly left a radium container in his vest pocket. It developed into skin erythema after 2 weeks followed by ulceration, which took several weeks to heal. Similar experience was reported by Pierre Curie with radium. It produced a burn on his forearm. Probably, this is the beginning of radiobiology.

Radiobiology is a scientific discipline that studies the effects of ionizing radiation on living systems including cells, normal tissues and malignant cells. It helps to understand the sequence of events such as the nature of exposure, biological damage, and its modification, and repair. The radiobiological studies contribute to the improvement of radiotherapy in many ways. Detailed knowledge of the human body is very much essential for the better understanding of radiobiology.

HUMAN ANATOMY

Human body begins with chemicals by the combination of atoms. The chemicals progress into cells, tissues, organs, systems and finally grown to organization level. At the chemical level, the atoms interact and form molecules or compounds with specific properties, e.g. water. The cells are made up of molecules, which are capable of performing precise chemical reactions.

The cell is a basic structure of human body and it carries out the vital functions such as metabolism, growth, irritability, adoptability, repair and reproduction. Cells are composed of atoms, which are bound to together to form molecules. Some molecules are arranged into small functional sources called organelles. Then organelles carry out specific functions within the cells, e.g. nucleus, mitochondria, endoplasmic reticulum, etc. These are classified according to their function namely:
- Muscle cells
- Bone cells
- Blood cells
- Nerve cells, etc.

Detailed study of cell will be discussed in the later paragraphs.

Tissues

Tissues are group of similar cells that perform specific functions, e.g. the muscle in the heart, pumps the blood. Various types of tissues are epithelial, connective, muscle and nervous. The percent composition of various tissues in human body is muscle (43%), fat (14%), organs (12%), skeleton (10%), blood (8%), subcutaneous tissue (6%), bone marrow (4%) and skin (3%). Epithelial tissue gives lining located inside the body and covering to outer surface of the body. These cells are further divided into squamous, cuboidal and columnar cells. Squamous cells are used for the formation of skin, cuboidal cells to lining the ducts (sweat glands) and columnar cells to line the digestive system. Connective tissue is the most abundant and composed of non-living material. It is divided into soft, fibrous, hard and liquid connective tissue. Its intercellular background is called matrix and the matrix may be semisolid, collagen, water or liquid.

Muscle tissue has elongated cells, which are highly cellular and vascularized. They help to produce movement, e.g. arm, heart, etc. The muscle tissue is divided into skeletal muscle, cardiac muscle and smooth muscle. However, the structure of the cells of the individual muscle is different from each other. Muscles can also be divided into voluntary muscles (e.g. skeletal muscle) and involuntary muscles (e.g. cardiac muscle). The voluntary muscle work on conscious, whenever there is a need, otherwise simply relax. On the other hand, cardiac involuntary muscle works on its own in the rhythmic manner.

Nerve tissue consists of neurons that are capable of generating and transmitting electrical impulses. Each neuron is made up of body, axon and dendrite parts, and its structure is similar to wire. The axon transmits impulses out of the cells, whereas the dendrites receive the impulses and transmit the signal to the cells of the body. Nerve tissue is responsible for the body functions such as feeling and movement.

Tissue Growth

Tissue grows basically in two ways, namely signal and repair. Growth by signal takes place during gestation period. The various tissues are formed at the end of the gestation period and reproduce with higher mitotic rate. Further, it grows with the help of growth hormones and attains physical characteristics. Growth by repair occurs during tissue injury by using regeneration or replacement. Regeneration has three steps namely inflammation, angiogenesis and proliferation. The injured cells release inflammatory chemicals that attract white blood cells (WBCs) and plasma cells to that area. Then, blood supply will be organized to the injured area by new blood vessel growth. Hence, nutrients and growth factors reach the area and tissue begins to proliferate. Instead of regeneration, fibrosis can also be used to repair injury in which the damaged tissue is replaced with a fibrous connective tissue. In that case, visible white scar may be present. Most of the tissues follow the above path except nerve tissue. Cells stop proliferation, when they begin to touch each other, called contact inhibition. The contact inhibition is missing in cancer cells.

Organs

Group of cells with similar structure form tissues and the different tissues bound together in turn form organs. Organ is usually made by two main tissues, namely parenchyma tissue and sporadic tissue. Parenchyma is the primary tissue, unique to the organ and only found in that organ. On the other hand, sporadic tissue is a secondary tissue found throughout the body. Each organ can have one or more primary tissues and several secondary tissues. In stomach, the epithelium is the primary tissue that performs secretion and absorption. On the other hand, connective tissue, vascular tissue, nervous and muscular tissues are secondary tissues.

Tissues and organs serve as independent unit and carry out specific functions. Cells in the tissue may be undifferentiated or differentiated depending upon the rate of proliferation and development. Undifferentiated cells are immature cells called stem cells, which are more radiosensitive. Cells over a period of growth and proliferation become mature and are called differentiated cells. Thus, radiosensitivity nature of the cell depends upon the state of maturity and its function.

Systems

Two or more functionally related organs working together is called system. The system carryout specific functions within the body. The human body systems are divided based on their functions, which are as follows:

1. **Integumentary system:** Thermoregulation and protection from environment.
2. **Skeletal system:** Support and protect different organs, mineral storage and blood formation.
3. **Muscular system:** Locomotion, support and heat production.
4. **Nervous system:** Responding to stimuli, control other organ activity.
5. **Endocrine system:** Regulates metabolism and body functions.
6. **Circulatory system:** Transport of oxygen, nutrients and removal of metabolic waste.
7. **Lymphatic system:** Drains excess fluid and provide immune response.
8. **Respiratory system:** Exchange of oxygen from air and carbon dioxide from blood.
9. **Digestive system:** Food processing, absorption of nutrients, minerals, vitamins and water.
10. **Urinary system:** Regulate volume and pH of blood, filter waste products.
11. **Reproductive system:** Production of eggs, growth of embryo and sperm.

Body Regions

The body regions are divided into head, neck, trunk, upper extremity and lower extremity. Trunk is further divided into thorax and abdomen. The head is having two regions, namely facial and cranial. The cranial region covers and supports the brain. The neck is called cervical region and supports the head and its movements. The chest is the thoracic region, which has two mammary regions laterally and a sternal region at the middle on the front. The armpit is called axilla and vertebral region follows the vertebral column on the back. The abdomen is below the thorax and has nine regions. The umbilicus on the front is the center of the abdomen. The pelvis region is the lower portion of the abdomen. The perineum region contains external sex organs and anal opening. The center of the backside is called lumbar region, followed by sacral region downwards. The large hip muscles form the gluteal region. The upper extremity consists of shoulder, arm, forearm and hand. The front and back of the hand is called palm and dorsum. The lower extremity consists of thigh, knee, leg and foot.

Life Maintenance

To maintain life, the body must be kept in a steady state known as homeostasis. This is mainly achieved by kidney, liver and brain organs. In addition, there is thermoregulation and osmoregulation. The former regulates the body temperature and keep it between 98 and 100°F (98.6°F). The latter controls the level of water and mineral salts in blood by osmotic pressure fluids. The heart rate is associated with inspiration rate and supply blood to all the parts of the body.

It exchanges carbon dioxide for oxygen in the lungs and oxygenates the blood. Increased or decreased heart rate is called tachycardia (> 100 beats per minute) and bradycardia (< 60 beats per minute). The system maintains the optimal heart rate of 60–90 beats per minute.

The kidney maintains the blood through water level, mineral content and pH value (7.4). It also filters waste materials such as urea and chemicals. The liver supplies energy to the body by storing glycogen. It maintains the blood sugar level with the help of pancreas. The brain is the overall control center that monitors the levels of glucose and salt molecule as well as waste materials. Without the brain, kidney and liver are unable to predict the status and requirements. Brain also has negative and positive feedback mechanisms. When the temperature rises, the body wants to cool, is an example for negative feedback. Childbirth is an example for positive feedback in which the cervix is stretched during the labor period.

CELL

Cell is the basic unit of life and human body is a multicellular organism. It is a structural and functional unit of every living organism. An organism can be either unicellular with one cell or multicellular with many cells. Human body is made up of 10^{14} cells and the cell size is 10 μm. The cell is formed with biomolecules and different substructures. There are two types of cells, namely prokaryotes and eukaryotes. Prokaryote is a primitive cell and there is no nucleus. Eukaryote is an evolved cell and has central core called nucleus, which is a membrane-bound organelle. Eukaryotes are further divided into somatic cell and germ cell. The germ cell is sperm in male and ovum in female, and it participates in the reproductive process. Every cell has four basic components, namely cell membrane, cytoplasm, cellular organelles and genetic material. In human body, there are 200 types of cells and each type is responsible for a specific function.

Cell Structure

Cell consists of a plasma membrane that encloses cytoplasm and nucleus (Fig. 1.1).

Plasma Membrane

The plasma membrane is a semipermeable sheet-like cover that contains lipid-protein complex and proteins. The membrane is constructed by bilayer of phospholipids. It selectively allows certain lipid-soluble substances to enter or leave the cell by diffusion. The substance includes nutrients and metabolic waste products. The cholesterol is interspersed throughout the membrane to strengthen the membrane. The proteins are embedded in the membrane that has several functions. The functions include formation of

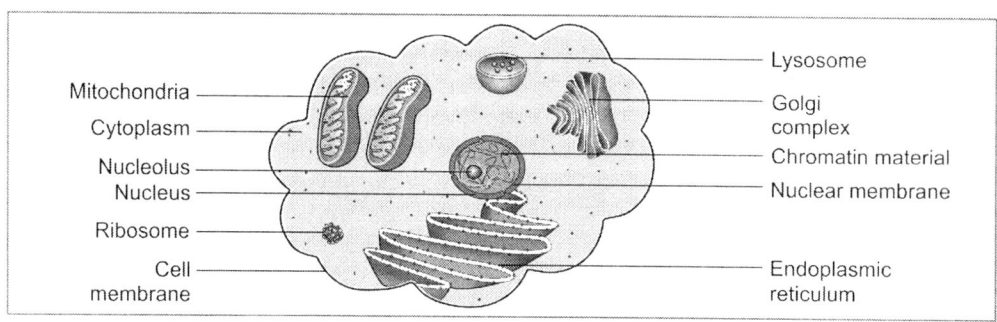

FIGURE 1.1: Cell structure

protein channels and receptor sites. The protein channels act as transporters and permit water and ions into the cell. They also move the molecules in and out of the cell. The receptors are in the outer boundary of the cell and transmit external signals to the cells. The plasma membrane provides structure and form to the cell.

Cytoplasm

The cytoplasm is an aqueous substance that lies between the outer cell membrane and the nucleus. It is made up of water, salts, biomolecules and other substructures, and occupies about 70% of the cell volume. The biomolecules are ribonucleic acid (RNA), proteins, carbohydrates and lipids. The substructures are organelles and inclusions. The organelles are:
- Endoplasmic reticulum
- Golgi complex
- Ribosome
- Mitochondria
- Lysosomes
- Microtubules.

The inclusions are either organic or inorganic compounds. Organic compounds contain carbon formed by living organism, proteins, carbohydrates and lipids. Inorganic compounds are water and electrolytes, which are not formed by living organisms. The chemical reactions such as glycolysis and cell division are taking place in cytoplasm.

Endoplasmic reticulum: It is a membranous structure-like tubes or flat sheet, or round chambers, which has network throughout the cytoplasm. It extends and connects the outer surface of the nuclear membrane. It is responsible for transport of materials and synthesis of few cellular constituents. It facilitates the communication between nucleus and cytoplasm. The endoplasmic reticulum may be rough or smooth. The rough endoplasmic reticulum is associated with minute organelles of RNA called ribosomes.

The smooth endoplasmic reticulum does not have ribosomes and involved in lipid, cholesterol and carbohydrate metabolism. They also do steroid hormone synthesis in testis and adrenals.

Ribosomes: These are tiny organelles attached to the outer surface of endoplasmic reticulum. They are large nucleoproteins containing 55–60 types of proteins and three to four RNAs. Each ribosome consists of 60% RNA and 40% proteins, and manufacture proteins using the information provided by the deoxyribonucleic acid (DNA). It serves as the site of protein synthesis and thus, help the cell function. There are two types of ribosome, namely free and fixed.

Golgi complex: It is made up of flat membrane disks called saccules. There may be five to six saccules that lie near the nucleus. Golgi complex mainly carries out secretory functions. They help to store and release biological secretions such as proteins, lipids and polysaccharides. Its major work is to synthesize carbohydrate-rich compounds such as polysaccharides and glycoproteins.

Mitochondria: It contains glycolytic enzymes that participate in cell metabolism. The main function is oxidation of food molecules such as proteins, carbohydrates and lipids. It synthesizes a special molecule called adenosine triphosphate (ATP). The ATP is rich in energy and participates in other metabolic processes and converts itself into adenosine diphosphate (ADP). Hence, mitochondria are known as power house of the cell. It also transports genetic information for synthesis of proteins.

Lysosomes: These are spherical structures and involve in the digestion process. It is richer in enzymes, which are enclosed by lysosomal membranes. It digests almost all chemical constituents of the cell and also controls intracellular contaminants. Malfunctioning of lysosome results in pathological disorder, which affects the cell's life. They are often called suicide bags.

Microtubules: These are protein-made structures and maintain cell shape.

Nucleus

The nucleus contains DNA, RNA, protein and water. DNA is the genetic material. Nucleus is the major constituent of the cell that is responsible for genetic information and initiate cell division. It is the largest organelle and acts as a control center of the cell. It is bounded with double membrane and is connected to the endoplasmic reticulum. This facilitates the passage of RNA-like molecules from nucleus to cytoplasm or vice versa. It also protects the DNA within the cell. The inner part is dense jelly-like mass that contains nuclear gel, chromatin and nucleolus. The nuclear gel contains proteins, DNA, RNA and phospholipids. DNA is responsible for encoding messages for all, including hair and color of the individual. The chromatin is a large fiber-like structure that is made up of DNA and basic proteins. Nucleolus is a spherical, membrane less structure and it participates in the transcription of genes, and contains enzymes required for RNA synthesis. There is no nucleus in red blood cells (RBCs), as they do not divide.

Chromosomes: These are multistranded structures, where the genetic material is placed. It is a thread-like structure containing DNA, which is a complex molecule. There are 23 pairs of chromosomes present in both sexes. Male have 22 pairs of chromosomes (autosomes) plus a pair of X and Y sex chromosomes. Female have 22 pair of chromosomes plus a pair of X and X sex chromosomes. Gene is a finite segment of DNA specified by exact sequence of bases. The bases are adenine, guanine, cytosine, thymine and uracil. Genes occurs along the chromosome in a linear fashion and its position is called locus. The 46 chromosomes contain about 6×10^9 base pairs of DNA. The total number of genes is about 30,000 per haploid set of chromosomes. An average gene may contain 30–60 kilo bases of DNA.

Cell Function

Cell has to do specific function to support the human body. In addition, it has to absorb nutrients and use them for energy production and molecular synthesis. If radiation exposure affects molecular synthesis, the cell may die or malfunction. Protein synthesis is an important cell function. The DNA in the nucleus contains the code that identifies the type of protein, to be synthesized. The code is determined by the base pair sequence. The base pair identify one of the amino acid out of 22 for protein synthesis.

Cell Cycle

There are two types of cells, namely somatic and genetic or gametic cells. The cell division in the somatic cells is called mitosis. The cell division in the genetic cells is called meiosis. Every somatic cell has four phases, namely mitosis (M), synthesis (S), gap 1 (G1) and gap 2 (G2). The later three phases are called interface, which is the period of cell growth (Fig. 1.2A). The typical cell cycle time is 10–40 hours of which G1 phase occupies 30%, S phase 50%, G2 phase 15% and M phase 5% of time. G1 time may vary and it is longer in slowly proliferating cell, whereas the cell cycle time in S, G2 and M phases vary little for different cells. The length of each phase in the cell cycle varies from cell line to cell line (Table 1.1). Mitosis is the shortest phase of the order of an hour, whereas S phase is about 6–8 hours. Renewing or growing cells participate in the cell cycle, e.g. skin, gut, bone marrow, tumor cells, cells in culture, etc.

Replication of genome occurs in S phase and mitotic propagation to daughter generations occurs in G2/M phases. Cells in the late G2 and M phase are most sensitive to radiation, whereas the cells in the late S phase are

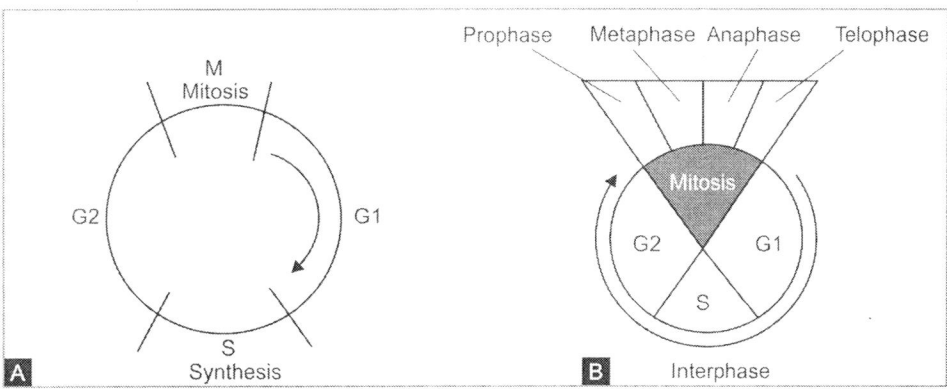

FIGURES 1.2A and B: Cell cycle of mammalian cells. **A.** Its various phases—M (mitosis), S (synthesis), G1 (gap 1) and G2 (gap 2); **B.** Subphases of mitosis in a cell cycle.

most resistant. Thus, variation of cell sensitivity depends upon the cell stage.

Table 1.1: Cell cycle and phase time for commonly used cells

Phases	Time for Hamster cells (h)	Time for HeLa cells (h)
M	1	1
S	6	8
G1	1	11
G2	3	4
Cell cycle	11	24

Recent studies have shown that the cell cycle progression is governed by protein kinases activated by cyclins. This is a periodic activation of different members of the cyclin-dependent kinase (Cdk) family. There are different Cdk-cyclin complexes that phosphorylate several protein substrates. These complexes not only drive the cell cycle events but also prevent wrong cycle events. The entry into S phase is regulated by cyclins D, E and A. The D-type cyclin acts as a growth factor sensor and its expression depends more on the extracellular cues, than on the cell's position in the cycle. Cyclin E expression in proliferating cells is periodic and maximum at G1/S transition.

Mitosis

The mitosis is the cell proliferation phase that consists of prophase, metaphase, anaphase and telophase (Fig. 1.2B). During the synthesis period, the DNA molecule is replicated into two daughter molecules. The chromosomes with two chromatids are transformed into four chromatids. Thus, two pairs of homogeneous chromatids with same DNA content and structure is formed (Fig. 1.3). Chromosome is not visible during the interface, whereas it is visible during mitosis.

Phases

Usually, chromosomes are long, thin and spiral shape.

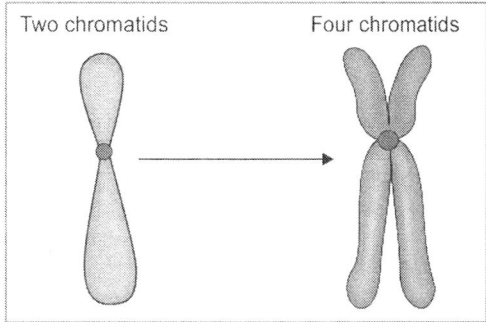

FIGURE 1.3: Chromosome replicate from two chromatids to four chromatids structure during the synthesis period

Prophase: During this phase, the nucleus swells and the cell condense the chromatin of the nucleus, resulting in the formation of highly structured chromosomes. Each chromosome has a slight staining constriction called centromere. The sister chromatids are attached to the centromere. The centrioles migrate to the pole of the cell. The chromosome attains maximum condensation at the end of the prophase and nuclear membrane disappears, resulting in the mixing of nucleoplasm and cytoplasm.

Metaphase: In this phase, the chromosomes move towards the center and line up along the equator of the nucleus. A spindle, composed of fibers, is formed, which connects the chromosomes to the pole. At the end of the metaphase, the chromosomes are stabilized at the equator. It is the suitable phase for chromosome analysis by an optical microscope. The mitosis can be stopped and studied for chromosomal damage, if any.

Anaphase: During this phase, the mitotic spindle gets shortened and the sister chromatids are pulled apart. As a result, the centromere undergoes degradation and separates the duplicated chromatids into two identical and distinct sets. They are pulled towards the poles by the microfibers attached to the spindle. Thus, complete division of DNA is achieved.

Telophase: In this phase, chromosome begins to uncoil and its structure disappears, and the nuclear membrane reappears. The nuclear membrane reforms around each set of chromatids and create two identical nuclei at opposite poles. In this phase, the cytoplasm and the organelles is divided into two equal parts to each nuclei, which is known as cytokinesis. This is followed by formation of furrow, which enlarges and results in two daughter cells. The dumbbell shape disappears and the cell division is completed. The two daughter cells are similar to that of the parent having the same genetic material.

Meiosis

Meiosis is the cell division that takes place in the genetic cells. The cell division is same and the subphases (prophase, metaphase, anaphase and telophase) are similar to that of mitosis. However, there are two divisions in genetic cells, namely meiosis I and meiosis II.

Genetic cell division phase

Each genetic cell produces two daughter cells and then undergoes second division. The M phase in meiosis has two phases, namely meiosis I (reduction division) and meiosis II (equational division).

Meiosis I: In this, the diploid cell chromosome number is reduced to that of haploid cell.

Meiosis II: In this, the chromosomes are partitioned equally among the daughter cells instead of replication. In meiosis II, there is no S phase and hence, no chromosome duplication takes place. Each phase of meiosis cycle has prophase, metaphase, anaphase and telophase, similar to mitosis cycle. At the end of meiosis II, there are four haploid cells with 23 chromosomes obtained from two haploid cells at the end of meiosis I. Thus, the chromosomes are reduced to half and hence, meiosis is the process of reduction division. During fertilization, the gametes ovum (female) and sperm (male) contribute 23 chromosomes each and making a total of 46 chromosomes.

Cell Cycle Checkpoints

To monitor the growth and DNA replication, there are checkpoints in the cell cycle. These are G1, G2 and metaphase checkpoints. The G1 checkpoint does the job of:

1. Monitoring cellular growth and environmental conditions.
2. Monitoring DNA damage.

If the cell does not have a particular size and protein synthesis, the cell cycle is ceased. It also monitors the DNA damage, if present, the cell cycle is halted and repairs are carried out. Then, only the cell is passed on to S phase. The G2 checkpoint monitors unreplicated DNA, replication error and damage within DNA. If required, cell cycle will be halted and repair will be carried out. After the completion of repair, cell will be moved to M phase. If the damage is too severe and repair is not possible, then apoptosis takes place. The mitotic checkpoint ensures the formation of complete mitotic spindle and chromosome alignment within the cell. The cell is given time to correct the defects, then only cell cycle is completed.

Binding of cyclins to cyclin-dependent kinases (Cdks) activates the kinase complex to negotiate the checkpoints. They are cyclin B/A and Cdk1 for G2/M transition, cyclin D1/Cdk4 for M/G1 transition, cyclin E/Cdk2 for G1/S and cyclin A/Cdk2 for S/G2 transition. The drugs capable of producing cell cycle blocks readily establish the B/A and Cdk1 pair, promote early mitotic progression before complete recovery and directly inhibits homologous recombination (HR) repair in G2, e.g. caffeine and pentoxifylline. In *p53* mutants and in cells of p53, null status arising from p53 destruction after viral infection, p21 induction may be abolished and p21-controlled inhibition of G1/S cannot occur. Hence, there is absence of G1 block and the cells enter the G2 as a block. Generally, tumor cells are *p53* mutant and shows altered checkpoint expression, and limited repair routes. This provides the opportunity for therapeutic intervention.

Cell Death

The life of a cell depends on the function of the tissue and the degree of cell damage. The cell death can be divided into autophagy, necrosis and apoptosis. Autophagy is a highly regulated process, which maintains balance between protein synthesis and degradation within cells. Starving cells mainly use the autophagy method for cell death. Necrosis is premature cell death caused by external injury. The injury can be traumatic or chemical, or radiation exposure. Necrotic cells are capable of releasing harmful chemicals to the surroundings, which can cause tissue damage. Apoptosis is programmed cell death; it is decided by the body. It involves series of events, so that:
- Its constituents can be reused
- No release of harmful chemicals to the surrounding tissue.

HUMAN BODY COMPOSITION

Human body is made up of cells, which are basic functional unit of life. A cell consists of the following molecules:
- Water
- Protein
- Lipids
- Carbohydrates
- Nucleic acid.

They are present in the human body in 80%, 15%, 2%, 1% and 1% respectively. The latter four molecules are said to be macromolecules, since they contain more than 100,000 atoms. Protein, lipids and carbohydrates are basically life-supporting organic molecules containing carbon. Nucleic acid is the most radiosensitive molecule. Small molecules can combine together and produce a larger molecule; process is called anabolism. Similarly, a large molecule can be broken into smaller molecules; process is called catabolism. Both anabolism and catabolism are generally called metabolism.

Ions

Ions such as hydrogen (H^+), sodium (Na^+), potassium (K^+), chlorine (Cl^-) and calcium (Ca^{2+}) are also present in the human body. Positive ions have excess protons than electrons and are called cations. The negative ions have excess electrons than protons and are called anions. Ions transport creates membrane potentials and permit the propagation of action potential in nerve cells. Living cells depend upon the H⁺ transport for the formation of ATP molecule.

Elements

There are more than one dozen elements in the body, but hydrogen (H), oxygen (O), carbon (C) and nitrogen (N) are the important ones. They account about 99% of the total number of atoms in the body. The percent composition of these elements is hydrogen (60%), oxygen (25.7%), carbon (10.7%), nitrogen (2.4%), calcium (0.2%), phosphorus (0.1%), sulfur (0.1%) and remainders (0.8%). The remainder includes elements such as potassium (K), sodium (Na), calcium (Ca), magnesium (Mg), chlorine (Cl), iron (Fe) and copper (Cu).

Molecules

Water

Water is a small and more abundant (66%) molecule in human body. It is about 80% in blood and kidney, and 10–20% in fat tissues. Majority of water is contained in the inner part of the cell. Human body is made up of structured water and it is available in free and bound state. It consists of two atoms of hydrogen and one atom of oxygen connected by intermolecular polar covalent bonds (Fig. 1.4). The oxygen has greater affinity to electrons; hence, electrons are usually nearer to oxygen atom. Two water molecules are joined by a hydrogen bond, which is weaker than the intermolecular covalent bond. However, collective hydrogen bond is stronger and is responsible for many physical properties of water such as high specific heat capacity, high melting and boiling points, and large surface tension.

Water is a stable molecule and greater amount of energy is required to split it. The specific heat of water is two times higher than that of other biological molecules.

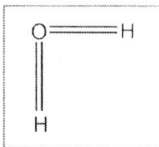

FIGURE 1.4: A water molecule with one atom of oxygen and two atoms of hydrogen connected by covalent bonds

Basically, water is a polar molecule having high dielectric constant. It is an effective solvent and all polar group molecules are soluble in water. Hence, they are called hydrophilic molecules, e.g. NaCl and KCl. Non-polar molecules are not soluble in water and hence, called hydrophobic molecules, e.g. triglycerides and aliphatic amino acids. Molecules with polar and non-polar nature are called amphipathic, which contain both hydrophilic and hydrophobic substances. The amphipathic phospholipids not only compartmentalize the cell but also maintain the cell as a highly organized biological system. The functions of water include:
- Dissolves big molecules and ions
- Maintains body temperature
- Maintains cell's shape
- Supplies energy to nucleic acid and helps metabolism.

Protein

Proteins are the molecules of biological importance and it accounts nearly 15% in the human body. There are more than 100,000 proteins in the human body. Protein molecule consists of carbon, hydrogen, oxygen, nitrogen and trace elements. Carbon, oxygen and nitrogen are the major atoms present in 50%, 20% and 17% respectively. The basic structure of protein is amino acid and there are about 20 essential amino acids in the human body. They are obtained from diet and used for protein synthesis. Amino acids are made up of a central carbon atom (C) attached to four chemical groups, namely:

- Amino group (NH_2)
- Carboxyl group (COOH)
- A hydrogen atom (H) and a variable residue called R.

The –R group distinguishes from one amino acid to the other and is responsible for various structural and functional properties of protein. They can be acidic, basic, hydrophobic or hydrophilic with complex ring structure or simple hydrogen atom.

Amino acids joined together with peptide bond or C—N bond (Fig. 1.5) in a specific sequence forms the primary structure. The secondary structure consists of alpha helices and beta-pleated sheets that form due to hydrogen bond. In the tertiary structure, a single protein can fold and form distinct geometrical shape due to hydrogen bond, disulfide bonds, electrostatic, ionic or hydrophobic interactions and van der Waals forces. The quaternary structure is simply the complexes of several small proteins.

Proteins provide structural support and help the transportation of other biological substances as they are high in muscles. In general, proteins are divided as globular and structural proteins based on their function. Globular protein functions as channels, enzymes, hormones (insulin), immune response (antibodies), intercellular transport (hemoglobin), intracellular transport (myoglobin), membrane bumps, membrane receptors and osmotic regulators (albumin). Structural proteins such as collagen, glycoprotein, microtubules and proteoglycans carry out specific function.

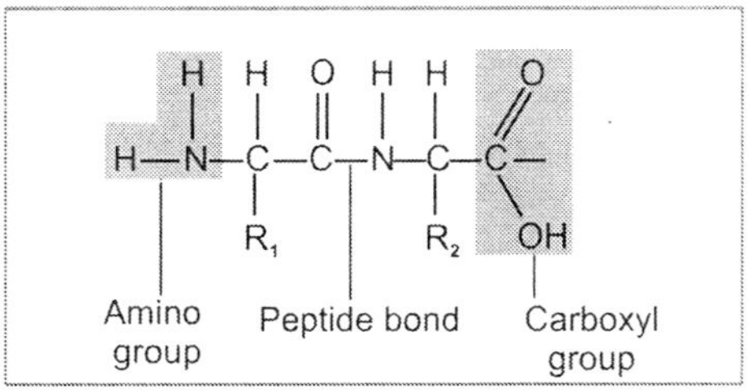

FIGURE 1.5: Primary structure of protein

For example, lipoproteins and hemoglobin protein transport lipids and oxygen respectively. Enzymes are molecules that permit biochemical reactions to continue, but do not participate in the reaction. Thus, enzymes act as catalysts and accelerate chemical process. There are two types of enzymes, namely apoenzyme and coenzyme.

Hormones are molecules that control body growth and development. It is secreted by pituitary, adrenal, thyroid, parathyroid glands, pancreas and gonads. Antibody is the primary defense mechanism, which work against infection and disease. It will have precise molecular design to tackle a particular infectious agent called antigen. Proteins can be denatured by excess heat, change of pH, salt, urea and organic solvents. They destroy electrostatic interactions, hydrogen bond, hydrophobic interactions and disulfide bonds, etc.

Lipids

Lipids are organic molecules of hydrophobic type consists of long hydrocarbon chains capable of forming impermeable barriers in aqueous solution. Lipids are made up of glycerol and fatty acid and each lipid consists of one molecule of glycerol and three molecules of fatty acid. The basic unit of lipids is fatty acid, which has a carboxylic group and a long hydrocarbon tail. There are two fatty acids, namely saturated and unsaturated. The saturated fatty acid is a non-polar type, has uniform electronic charge distribution without carbon-carbon double bond. The unsaturated fatty acid is a polar type, has one or more carbon-carbon double bond and asymmetric in charge distribution. For example, three fatty acids combine to form triglyceride, which does the function of storing energy.

Lipids are digested by the action of pancreatic lipases in the small intestine and bile acids that form micelles. They transport the lipids to the intestinal cell wall, where it is absorbed in the jejunum of small intestine.

From the intestinal lumen, lipids are transported to the lymphatic system through lacteals. Finally, it is drained into the thoracic duct, which joins the circulatory system in the upper left chest. Now, it is distributed to tissues or transported to the liver for further processing. It is present in all tissues and cell membranes. It is present in the skin and acts as a thermal insulator. They support membrane structure and store chemical energy for metabolism.

Carbohydrates

Carbohydrate is abundant in human cell and consists of carbon, hydrogen and oxygen. The number of hydrogen to oxygen ratio is 2:1, similar to that of water. It gives structure, shape, stability, energy and exists in the form of monosaccharide, disaccharides and polysaccharides. Monosaccharide is a sugar molecule that contains 3–10 carbon atoms, e.g. glucose, galactose, pentose and hexose. Two monosaccharides get condensed and forms disaccharide, e.g. sucrose, maltose, cellobiose and lactose. Polysaccharides contain more than 10 monosaccharides linked by glycoside bonds, e.g. starch, amylopectin and amylose. If all the monosaccharides are of the same type, it is called homopolysaccharides, e.g. cellulose. If it contains different types of monosaccharide, it is called heteropolysaccharides, e.g. hyaluronic acid.

Sugars are generally described by the formula $C_n(H_2O)_n$. Carbohydrates provide fuel for cell metabolism. Monosaccharides are transported from the gastrointestinal lumen into the blood with the help of membrane transporters. Polysaccharide (glycogen) stored in the tissues are easily converted into glucose for energy. This option is used, if the regular sugar and glucose is insufficient in the body. The blood glucose levels are maintained by various metabolic processes including glycolysis, gluconeogenesis, glycogen synthesis and glycogenolysis. High level of glucose may lead to excess glycosylation of proteins in the body and insufficient level will reduce the fuel to the brain. There are non-digestible carbohydrates that include plant sugars such as cellulose, pectins and lignins.

Nucleic Acid

Nucleic acid is a constituent of genetic material and is a largest macromolecule. It is responsible for protein synthesis in cells. There are two types of nucleic acid, namely deoxyribonucleic acid (DNA) and ribonucleic acid (RNA). The DNA is found in chromosomes and mitochondria, whereas, RNA is present throughout the cell. There are three types of RNA, namely:
- Ribosomal RNA (rRNA)
- Messenger RNA (mRNA)
- Transfer RNA (tRNA).

They participate in protein synthesis and help cell growth and development. The amount of RNA and DNA in a cell vary with cell type and function.

DNA and RNA

The DNA and RNA are polymers arranged in certain sequence with four different nucleotides. Each nucleotide consists of carbon, oxygen, hydrogen, nitrogen and phosphorus, which forms the three important basic molecules such as nucleobase, carbohydrate and phosphorus, alternatively known as nitrogenous base, sugar and phosphate group.

Structure

The nucleobase is formed by nitrogen and carbon atoms responsible for pairing of DNA polymers. In the DNA strand, they are bound together by hydrogen bond, which is weak and easily broken. The nucleobases/nitrogenous bases are adenine (A), guanine (G), cytosine (C), thymine (T) and uracil (U). The first two bases such as A and G are called purine bases, and C, T and U are called pyrimidine bases. DNA consists of A, G, C and T, whereas RNA consists of A, G, C and U. Uracil occupies the place of thymine in the case of RNA (Figs 1.6A and B).

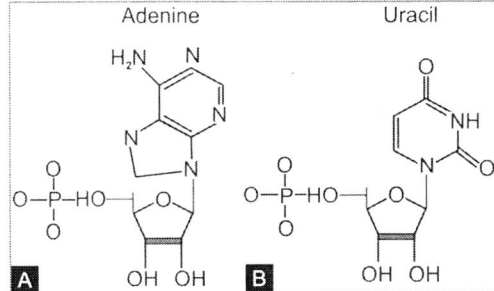

FIGURES 1.6A and B: Nucleotides. **A.** Deoxyribonucleic acid nucleotide of adenine base with deoxyribose as carbohydrate; **B.** Ribonucleic acid nucleotide of uracil base with ribose as the carbohydrate. Both has phosphate group attached to the carbohydrate.

Each nucleobase is bound to a carbohydrate and a phosphate group. The carbohydrate is pentose sugar and the phosphate is a single phosphate atom surrounded by minimum four oxygen atoms. The sugar in RNA is ribose and that in DNA is deoxyribose. The nucleobase added with above carbohydrate and phosphate is called nucleotide, which is the basic building block of DNA and RNA. Nucleotides are joined together with ester and phosphodiester bonds, which are stronger than hydrogen bond. The ester bond is flexible and binds the sugar to the nucleobase at the DNA strand. Now, phosphate group is connected to the sugar by the phosphodiester bond, which is a strong covalent bond. DNA is the most radiosensitive molecule and has alternate sugar and phosphate group. Due to the asymmetric nature of phosphate group, DNA and RNA molecules are given direction as determined by the terminal end. If the terminal end is a phosphate group, it is labeled as 5' (five carbon of sugar). If it ends with sugar group at the terminal, it is labeled as 3' (three carbon of sugar). This rope-like structure is called DNA polymer.

This is the primary structure of DNA and the secondary structure is achieved by complementary DNA strand (Figs 1.7A and B). The side of the strand is very critical that has alternate carbohydrate and phosphate groups. The polymers occur in the opposite directions of the strand and hydrogen bond is formed between the nucleotides. Thus, human DNA has two long chains attached together. This looks similar to a ladder, twisted around a common axis and looks as a double helix model. Each sugar molecule is attached with an organic base namely thymine, adenine, cytosine and guanine. Two hydrogen bonds are formed between thymine and adenine (T—A), whereas there are three hydrogen bonds between cytosine and guanine (C—G). The bases are paired similar to this along the length of DNA molecule and the T—A and C—G appears as rung on a ladder. During the S phase, DNA synthesis take place and each base pair is precisely duplicated.

The sugar-phosphate unit is outside the helix, whereas the bases are inside the helix. Each base pair is added perpendicular to the long axis of the DNA polymer with a spacing of 0.34 nm. The double helix DNA takes further twists and turns (packaging) in space and gives the tertiary structure, i.e. the DNA is first coiled around a series of histones, which are set of eight proteins called octamer (Figs 1.8A to D). The DNA is winded around the histone surface in a helical pathway.

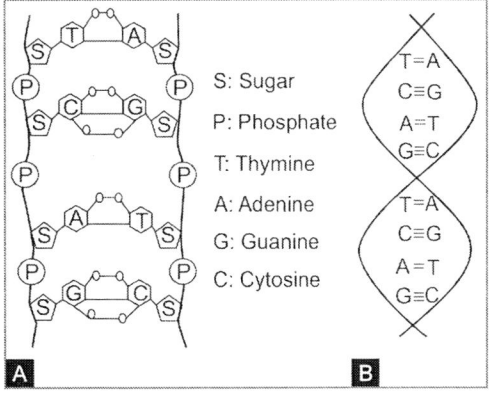

FIGURES 1.7A and B: Deoxyribonucleic acid (DNA) molecule. **A.** Made up of two long chains (ladder) of alternating sugar and phosphate molecules with pairs of base by bonding; **B.** DNA ladder twisted about an imaginary axis and form double helix.

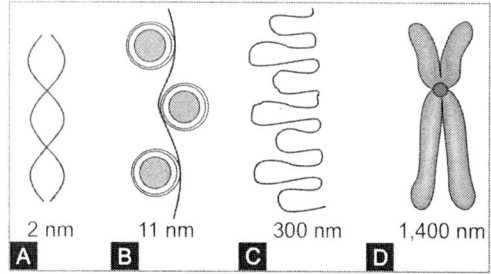

FIGURES 1.8A to D: Various stages of folding and packing of deoxyribonucleic acid (DNA). **A.** DNA double helix form; **B.** Chromatin—'beads on string' form; **C.** Chromosome in extended form; **D.** Chromosome at the metaphase.

Two complete winding of DNA may have about 150 base pairs. In this stage, the histone complex is called nucleosome (2.5 nm), which is similar to a string of beads. This make the DNA strand well protected, but prevents the accessibility of DNA to transcription and regulatory proteins. Hence, the DNA unwinds from the histones to provide more access for above such tasks. Double strand break is not possible, since the strands are not exposed. The nucleosome is further condensed and form chromatin structure resulting in chromosome, which is visible during mitosis. The total length of DNA is about 2 m, which is condensed into a nucleus of about 6 μm diameter.

By heating the DNA, the hydrogen bonds can be broken and it is called DNA melting. This can be measured in the laboratory by estimating the light absorption.

Functions of DNA

The function of DNA is to carry genetic information, which is used during organism development. DNA polymer contains both gene and nongenes, and their separation depends on the sequence at a particular location. Thus, chromosome carries the information of an individual characteristic in a coded form. The gene is a finite segment of DNA specified by an exact sequence of bases and occurs along the chromosomes in a linear order resembling 'beads on a string'. The gene contains instructions for protein synthesis and its position is called locus. The method of moving the above instructions out of the nucleus is called transcription. For transcription, a complementary RNA copy of a specific DNA is obtained. Since RNA is a single strand, only one strand of DNA is copied. RNA is single polynucleotide chain and their secondary and tertiary structure does not follow any regular pattern. The RNA synthesis takes place in the following steps:

- The DNA unwinds and allows access to nucleotides
- RNA and DNA nucleotides pairing takes place, thymine is replaced with uracil base
- Formation of RNA polymer with addition of nucleotides and alternating carbohydrates and phosphate molecules
- Hydrogen bond between nucleotides of DNA and RNA breaks, and synthesized RNA is freed
- The RNA moves out of the nucleus with information copied from DNA to ribosomes, which is a site of protein synthesis.

The ribosome decodes the RNA and produce specific amino acid chain for protein synthesis, which is called translation. The translation involves steps of activation, initiation, elongation and termination. The genetic code decides how the DNA sequences are translated. The genetic codes are set of rules that contain instructions, which are coded in the DNA. As per the code, three nucleotides can be read at a time, which is called codon. The start codon is set of three red plates and the terminal codon is the set of three blue plates. Once translation is over, the protein can be used by the organism. Only 20 amino acids are used for each protein. Since four nucleotides are read in groups of three, there are only 64 (4^3) possible codes.

Genome

Genome is composed of DNA of chromosome and DNA of mitochondria. Its integrity should be maintained in order to maintain the genetic accuracy. A genome may face more than million DNA breaks and lesions in every day. However, there are mechanisms to repair the above damage. In addition, there are checkpoints that are located at G1 and G2 phases. The G1 checkpoint locates DNA breaks, which is repaired during the cell cycle. In the G2 checkpoint, the DNA is reviewed for mismatched nucleotides and unreplicated DNA. If necessary, the cell cycle is paused and repair is again carried out. Thus, the genome integrity is maintained.

If the breaks and lesions are not repaired or integrity of the genome is compromised, mutations may occur. The mutation denotes change in the chromosome, their genes and their DNA. Thus, genome instability is caused by chromosomal alterations. It can lead to deletion or insertion of DNA or change in chromosome number. If the daughter cell receives more than one copy of a chromosome, the cell is called aneuploid, which is an abnormal number of chromosomes in a cell. It is common in chromosome 21 in humans that too in developing fetus resulting in Down syndrome. It copies three chromosomes instead of two. Mutation takes place both in germ cells and somatic cells. Mutation can cause hereditary diseases that are transmitted to progeny.

MOLECULAR FORCES AND BONDS

Biomolecules usually have multiple structures and their primary structure is a stable one. The structure arises out of interactions between their constituent atoms. The structure can be described with the existence of strong and weak forces.

Strong Forces

A strong force is a short range attractive force responsible for covalent bonds between atoms. It is responsible for the primary structure of biomolecules. Covalent bond is the coupling between electrons with opposite spins. Hence, the electrostatic repulsion has overcome and results in an attractive force called covalent bond. Basically, the electrons shared between two atoms in a covalent bond form a cloud of negative charge and balances the repulsive forces. The bond length is the distance between two atoms at which the repulsive force is nullified. Oxygen has six valence electrons; can form two covalent bonds with hydrogen, to complete the outer electron shell. Similarly, carbon has four valence electrons, which can form four covalent bonds with other atoms. The bond between sugar and phosphate molecule in DNA is a covalent bond.

Bonds can be formed in between ions, which are either positively or negatively charged. This type of bond is called ionic covalent bond or simply ionic bond. It is a result of atoms striving to attain a full outer shell of electrons. The ionic bond is differentiated from the covalent bond with dipole moment. If the dipole moment is less than 0.1 OD, then the bond refers pure covalent bond. These bonds are easily dissociated in water. Both covalent and ionic bond exists in atoms that have binding energy of few electron volt (eV).

Weak Forces

The weak force refers long range forces between molecules that give them secondary structures. It makes the molecule more flexible, mobile and performs different functions. This type of force exists between protein and carbohydrates, fats and nucleic acids. The types of weak forces are:
- Coulomb force
- Charge-dipole force
- van der Waals force.

Both coulomb and charge-dipole forces can be applied to ionic bond also.

Coulomb Force

The Coulomb force is the electric interaction between two point charges separated by a distance 'r' and is proportional to $1/r^2$.

Charge-dipole Force

The charge-dipole force is the interaction between two opposite charges separated by a distance 'r' and is proportional to $1/r^3$.

van der Waals Force

Atoms and molecule has a weak attractive force between each other that is responsible for the condensation of gases into liquids and freezing of liquids into solids even in the absence of ionic or covalent bond. The above force is called van der Waals force. It is a short range force between molecules proportional to $1/r^7$, where 'r' is the distance between them. It determines the bulk properties of matter such as surface tension, viscosity, etc. These forces do not involve electrons, but are cumulative in nature. The types of van der Waals forces are:
- Dipole-dipole force
- Inductive force
- Dispersion force.

Hydrogen bond: This is a type of van der Waals force that occurs between certain molecules containing hydrogen atom. It is weaker than covalent and ionic bonds, but stronger than the normal van der Waal coupling. It belongs to medium strong force. The bond lengths are longer than covalent bond and smaller than the sum of the van der Waals radii. This is common in large biomolecules that are responsible for their secondary structures. It is usually formed between two electronegative atoms (e.g. O, N), one atom is called acceptor and the other atom is called donor. The acceptor and donor atoms are 'F, O, N' and 'F, O, N, Cl' respectively. Hydrogen bonds are directional and perform polar interaction. The polar interaction occurs between an electropositive hydrogen atom and two electronegative atoms, but forms covalent bond with hydrogen. Materials with hydrogen bond have high values of heat of vaporization, melting point, boiling point and dielectric constants. Hydrogen bond determines the structure of water.

TUMOR BIOLOGY

Tissue normally has regulated cell division and self-elimination capacity, which is a programmed cell death. This is very much required to achieve growth, which is a process of increase in size and maturity of cells from fetus to adult.

Important Concepts

Growth Process

Growth process consists of proliferation, differentiation and apoptosis.

Proliferation: It is the ability to divide.

Differentiation: It is a specialization in function.

Apoptosis: It is a programmed cell death, which eliminates unwanted or defective cells.

Growth Disorder Factors

There are growth disorder factors such as hypertrophy, hyperplasia, metaplasia and neoplasia as detailed below.

Hypertrophy: It is the increase in size of the organ due to increase in its cells.

Hyperplasia: It is the increase in size due to increase in the number of cells, which is reversible.

Metaplasia: It is a change of tissue from one form to another.

Neoplasia: It is a new growth, irreversible and called cancer.

Neoplasm

Neoplasm can be defined as a lesion resulting from abnormal growth of cells, which may be autonomous or relatively autonomous. Neoplasm can be either benign or malignant, cells or tumor.

Benign tumor: This remains locally, grow slowly, has encapsulated margin and good differentiation.

Malignant tumors: These are locally destructive, rapidly growing or variable growth, invasive margin, variable or poorly differentiated and capable of spreading to other parts of the body (metastases).

Neoplasm can be broadly classified as:
- Carcinoma
- Sarcoma
- Lymphoma.

Carcinoma: It is a Greek word referring crab, originates from epithelium and accounts about 75% of the malignancy. It is further divided into squamous cell carcinoma (mouth, pharynx, larynx, esophagus, cervix and vagina) and glandular carcinoma (secretory glands, endocrine glands).

Sarcoma: It arises from mesenchymal cells.

Lymphomas: These are from lymphoid cells.

Mutation and Genes

The uncontrolled cell division and failure of self-elimination causes tumor or cancer. Thus, malignant transformation occurs through accumulation of mutations in a specific cell. The reason for the above deregulation is the alterations of three groups of genes, namely:
- Proto-oncogene *(Ras)*
- Tumor suppressor genes *(p53)*
- DNA stability genes.

Proto-oncogene: It codes the protein that stimulates cell division. The proto-oncogene can be activated into oncogene by retroviral integration, point mutation, gene amplification and chromosome rearrangement. The oncogenes are either mutant or abnormally expressed form, which has increased transcription, protein and activity. Thus, oncogene can result from a mutation, deletion or alteration in the expression of one copy of a gene.

Tumor suppressor gene: It suppresses the effect of oncogenes on transformation and tumor formation. They oppose the biochemical functions of proto-oncogenes. Inactivation of both copies of tumor suppressor genes is required for loss of function.

Proto-oncogene and tumor suppressor genes are the components of the signaling network; the former acts as a positive growth regulator that can activate oncogenes, whereas the latter acts as negative growth regulator that can oppose the oncogene.

DNA stability gene: It monitors and maintains DNA integrity. Any loss of the gene may lead to defective sensing of DNA lesions and improper repair of the damaged template.

Development of Cancer

Cancer is outcome of series of gene mutations.

Gene Mutation Hypothesis

There are three hypotheses on gene mutations, leading to cancer. They are:
- Random multimutation
- Master gene
- Multichromosomes.

Random multimutation: This assumes that the mutations are random, follow no order and the first mutation is pertaining to metastases.

Master gene: This approach says there are few master genes that control the growth rate of cells. If any one of the master gene gets mutated, the regulation of growth rate will be lost.

Multichromosomes: The chromosome hypothesis believes that cancer cells have 60–90 chromosomes, instead of 46 or 23 pairs of diploid form. This creates stress such as DNA break, deletions and substitutions during the reproduction. The additional chromosomes are found to be aneuploid cells. However, the exact mechanism of cancer is not well understood and the following is the current understanding of cancer.

A normal tissue progresses into a tumor in various steps over a period of time. These arise from mutations, deletions or change of genes. Mutations activate oncogenes, inactivate tumor suppressor genes and reduce the activity of DNA stability genes. Subsequently, the cells undergo immortalization, transformation and metastasis. Thus, cancer formation is a multistep process starting from neoplasm stage to metastasis stage. It is derived from a single cell and is characterized by unrestrained growth, irregular migration and genetic diversity. There is no threshold for the induction of cancer, but the probability of cancer induction is generally low due to inherent defense mechanisms. However, the complete etiology of cancer is not well defined. In general, the individual's diet, lifestyle, genetic and environmental conditions are the important factors of cancer risk. Especially, tobacco, alcohol, ultraviolet, ionizing radiation and pollution are few agents that increase cancer risk.

Steps in Cancer Development

In clinical practice, the multistep process of cancer (Fig. 1.9) is termed as:
- Initiation
- Promotion
- Progression.

Initiation: Any damage occurs in normal cell, gets repaired and performs its normal function. If it is repaired with some mistake, it may lead to mutation. The cell is viable, but do not perform the normal function of the cell and it is called initiation. This is the first step towards neoplasia and may be due to change in molecular level and function. The mutated cell forms precancer stage, which is not clinically observable. The agents that create initial neoplasia are called carcinogens. In addition, agents such as tobacco, ionizing radiation, chemicals, asbestos, hormones and viruses can also cause the above initiation stage.

Promotion: In this stage, the other agents such as physical, chemical or viral may promote the preclinical stage to precancer stage called minimal cancer, which also has no clinical sign. During this stage, the preneoplastic cell is stimulated to divide. The promoter need not be a cancer-causing agent, once cancer is initiated in a cell, it promote that cell further into precancer stage.

Progression: This is a final stage in which cancer-inducing agents may convert the precancerous cells into cancerous with clinically observable signs. In this, the transformed cells produce a number of phenotype clones. All the phenotypes are not the

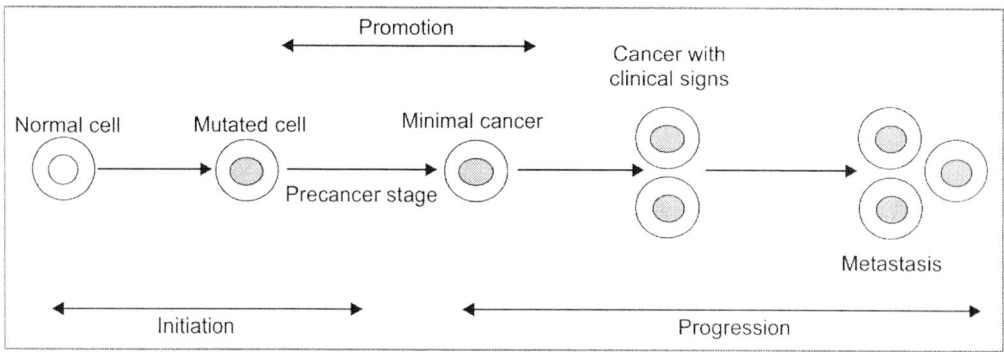

FIGURE 1.9: Development of cancer and its stages from a normal cell

neoplastic type; there may be one, which evade the defense mechanism and becomes cancer. Cancerous cell may become metastasis malignant cells, which may spread to other organs through lymph and blood vessels.

A single genetic alteration is not sufficient to cause tumor. Multiple genetic alteration occurring over a period of time forms solid tumor. Most of the alterations are due to somatic mutations, which include chromosomal translocations, deletions, inversions, amplifications or point mutations. Such mutations in proto-oncogene and tumor suppressor genes permit selective growth of premalignant cell that attains colonial status. On the other hand, the mutations in DNA stability genes increase the rate of genetic mutations, which results in malignant tumor.

Most of the malignant tumor cells are heterogeneous in population and fail to form tumor. However, few tumor cells have the ability to repopulate and form tumor. Hence, these types of tumor cells are said to have stem cells. The difference between the normal and malignant cell is shown in Figures 1.10A and B, and summarized in Table 1.2.

Table 1.2: Difference between normal and cancer cells

Normal cells	Cancer cells
Exhibit genetic stability, which is important for integrity of the organism and maintain the function of tissues	Exhibit genetic instability, it may be either microsatellite instability or chromosomal instability
Cells die by apoptosis or necrosis	Evade apoptosis
Require external growth signal to begin the reproductive cycle	Ignore antigrowth signals, generate their own internal growth signals
Obey antigrowth signals at tissue borders	Ignore antigrowth signals at boundaries
Have their own blood supply	Generate their own blood supply
Cells have about 60 reproductive cycles and cycles are limited	Cells produce indefinitely and have ability to metastasize
Ability to maintain their differentiation, retain tissue function	Lose their differentiation, unable to maintain their tissue type

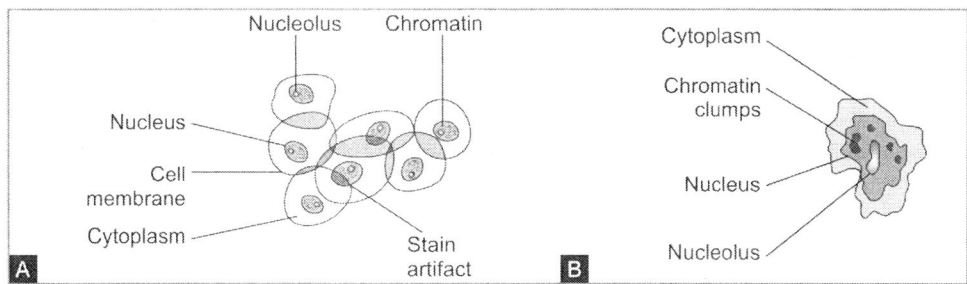

FIGURES 1.10A and B: Normal and malignant cell. **A.** Cluster of normal cells with large amount of cytoplasm with clear nuclear stain, fine chromatin granules, small nucleolus, smooth nuclear border and stain artifact; **B.** Malignant cell with large nucleus with dark staining, large chromatin clumps, large nucleolus and irregular nuclear border.

Cancer Staging

Staging is the estimation of extent of the tumor spread, which influences type of treatment and prognosis. Staging includes clinical, pathological, radiological and biochemical information.

Basic Classification

The basic classification of staging is:
- Stage 1: Tumor confirmed to the organ of origin.
- Stage 2: Invade local lymph nodes.
- Stage 3: Distant nodes invaded or local spread beyond the origin.
- Stage 4: Blood-prone metastasis present.

TNM Classification

The TNM classification Union for International Cancer Control (UICC) describes the primary tumor (T), nodal spread (N) and distant metastases (M). T0 refers that there is no primary tumor; T1, T2, T3 and T4 refers tumors of increasing size, and T4 indicate the most advanced local disease with invasion to adjacent structures. N refers the lymph nodes and their number; N1 indicates the mobile nodes on the primary side, N2 indicates mobile node on the opposite side of the primary and N3 refers fixed involved nodes. The node is either positive or negative. The ratio of the negative to positive node is of prognostic importance. The M indicates the presence (M1) and absence (M0) of metastases at distant sites.

Other Classifications

In addition, there are American Joint Committee on Cancer (AJCC) and Federation Internationale de Gynecologie et d'Obstetrique (FIGO) staging systems in practice.

In clinical practice, cancer is often graded as well differentiated, moderately differentiated, poorly differentiated and undifferentiated. For example, the well differentiated is lower grade, since it looks similar to a normal cell, except a little difference. Undifferentiated cells are higher grade, since they are more primitive, reproduce faster and have higher mitotic index.

BIBLIOGRAPHY

1. Ann Barrett, Jane Dobbs, Stephen Morris, et al. Practical Radiotherapy Planning, 4th edition. London: Hodder Arnold; 2009.
2. Charles A Kelsey, Philip H Heintz, Danial J Sandoval, et al. Radiation Biology of Medical Imaging. Hoboken, New Jersey: Wiley Blackwell; 2014.
3. James Claycomb, Jonathan Quoc P Tran. Introductory Biophysics: Perspectives on the Living State. New Delhi: Jones & Bartlett; 2011.
4. Sampath Madhyastha. Manipal Manual of Anatomy for Allied Health Science Courses, 2nd edition. New Delhi: CBS Publishers & Distributors Pvt Ltd; 2007.
5. Srivastava PK. Elementary Biophysics: An introduction. New Delhi: Narosa Publishing House (P) Ltd; 2005.

Radiation Physics

ATOMIC STRUCTURE

Matter is composed of elements and compounds. Elements are the simplest chemical entity, which cannot be broken further, e.g. hydrogen, carbon, oxygen, etc. Two or three elements form a compound, e.g. water. The smallest particle of an element is the atom, which forms the fundamental unit of matter. Atoms are very small and the diameter is of the order of 10^{-10} m. Every atom possesses a central core called nucleus, which is positively charged. The diameter of the nucleus is of the order of 10^{-14} m.

The nucleus consists of two kinds of nuclear particles called protons and neutrons, and collectively known as nucleons. The proton is positively charged and the neutron has no charge. The space around the nucleus consists of another important particle called electron. The electrons are negatively charged particles and they circulate around the nucleus at varying distances, similar to planets rotating around the sun (Fig. 2.1). The number of electrons in an atom is equal to the number of protons, hence atom is said to be neutral.

There are two types of forces existing in the nucleus. The electrostatic repulsive force, exist between particles of similar charge. The strong forces (attractive) resulting from the exchange of pions among all nucleons, hold the nucleus together. These two forces act in opposite directions. The nucleus has energy level and the lowest energy state is called ground state. Nuclei with energy excess of the ground state are said to be in an excited state. Excited states that exists $>10^{-12}$ s are referred as metastable or isomeric states.

Recent studies have shown that there are hundreds of elementary particles in the atom. Proton and neutrons are made up of quarks, and there are six types of quarks namely, up-, down-, strange-, charmed-, bottom- and top-quarks. The other elementary particle includes photons, bosons, baryons, leptons, mesons, gluons, muons and neutrinos.

Important Aspects

Atomic Number and Mass Number

In 1913, Mosley HGJ stated that the atomic number of an atom is the number of protons in the nucleus. It is also equal to the number of electrons of the atom, which is represented by Z. The mass number of an atom is the total

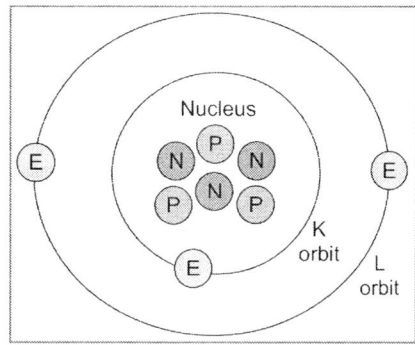

FIGURE 2.1: Atomic structure (E, electron; P, proton; N, neutron)

number of protons and neutrons in the nucleus, and it is denoted by A. An element (X) is symbolically described as $_Z^A X$. The subscript gives the atomic number Z, while superscript gives the mass number A.

Isotopes

The atoms composed of nuclei with the same number of protons, but with different number of neutrons are called isotopes. In other words, isotopes have the same atomic number and different mass numbers, e.g. hydrogen have three isotopes namely:
- $_1^1 H$, has 1 proton (hydrogen)
- $_1^2 H$, has 1 proton and 1 neutron called deuterium
- $_1^3 H$, has 1 proton and 2 neutrons called tritium.

Isotopes of an element have the same chemical properties, but have different physical properties. Isotopes capable of performing radioactivity and emitting radiations are called radioisotopes, and their nuclei are said to be unstable. Nuclides having the same mass numbers, but different number of protons are called isobars. Nuclides having the same number of neutrons, but different number of protons are called isotones. An isomer is the excited state of a nucleus and it will have same number of proton and neutron.

Electron Shells

In 1921, Burry and Bohr independently gave a scheme for the arrangement of electrons in an atom. According to this scheme, the orbits in the atom are named as shells and denoted as K, L, M, N, etc. from the nucleus. The following are the rules of their scheme:
1. The maximum number of electron in each shell can be obtained from formula $2n^2$, where n = 1, 2, 3, 4, etc. In the case of K shell n = 1, the number of electrons in the K shell = $2 \times 1^2 = 2$.
2. In the case of L shell n = 2, the number of electrons in the L shell = $2 \times 2^2 = 8$ and so on.
3. Each shell is provided with subshells, which are denoted as s, p, d, f, etc.
4. The K shell (n = 1) has one subshell, namely 1s.
5. The L shell (n = 2) has two subshells, namely 2s and 2p and so on.
6. One electron in the s subshell of K shell is denoted as $1s^1$, while 2 electrons in the same subshell are denoted as $1s^2$.

The outermost orbit is called valence shell, which is responsible for chemical, thermal, optical and electrical properties of the element. No valence shell has more than 8 electrons, e.g. metals have one, two or three valence electrons. The elements are arranged in the periodic table based on the similarities of chemical properties of different elements. As we go across the periodic table the atomic number of the atom increases. The number of electron also increases in the same step.

Quantum Number

The energy level of an electron or position in an atom is described by quantum numbers as follows:
1. The principle quantum number (n) defines the main energy level or shell of an orbiting electron, e.g. K shell, n = 1; for L shell, n = 2 and so on.
2. The azimuthal quantum number (l) describes the angular momentum of the orbiting electrons. It can have values 0, 1, 2, 3 .... n – 1, e.g. the M shell principal quantum number is 3 and its azimuthal quantum numbers are 3 – 1 = 2, which are 0, 1 or 2.
3. The magnetic quantum number (m) describes the spatial orientation of the plane of the orbiting electron and it can have values from –l to +l. When l = 1, m can have –1, 0, +1 values.

4. The spin quantum number(s) describes direction of spin of the electron and it can value +½ (spin up) or −½ (spin down).

Ionization

An atom is normally electrically neutral. If one or more orbital electrons are removed from the atom, the remainder of the atom is left with positively charged and is known as positive ion. This process of removal of orbital electrons from the neutral atom is known as ionization. Sometime one or more electrons may attach themselves to a neutral atom and form a negative ion. This is also known as ionization.

Binding Energy

To produce ionization, energy must be given to an orbital electron in order to remove it from the atom. There is a force of attraction between the electron and the positively charged nucleus. Therefore, the energy that is just sufficient to remove an electron from the orbit is known as binding energy. The magnitude of the binding energy depends on the atomic number and the shell from which the electron is being removed. It is greater for elements of higher atomic number and greatest for the K shell (inner most shell).

Binding energies are negative because they represent amounts of energy that must be supplied to remove electrons from atoms. Electron shells are often described in terms of the binding energy of electrons occupying the shells, e.g. the binding energy is −13.5 eV for an electron in the K shell of hydrogen and −3.4 eV for an electron in the L shell. The K shell-binding energies of various elements are given in Table 2.1.

Excitation

In an atom, if energy is supplied, the electrons can be moved from the inner orbits to the outer orbits. Now, the atom will have more energy than its normal state. It is said to be in an excited state and the process is known as excitation. For example, to move an electron from K to L shell of the hydrogen atom, the energy required is:

(−3.4 eV) − (−13.5 eV) = 10.1 eV

Table 2.1: Atomic number (Z) and binding energies (E_k) of few elements

Element	Z	E_k (keV)
Aluminum	13	1.6
Calcium	20	4
Molybdenum	42	20
Iodine	53	33
Barium	56	37
Gadolinium	64	50
Tungsten	74	70
Lead	82	88

Electron Volt

When energy is measured in the macroscopic level, units such as Joule (J) and kilowatt-hours (kWh) are used. In the microscopic level, the electron volt is a more convenient unit of energy. Hence, electron volt (eV) is used as the unit of energy in radiation physics. One electron volt is the kinetic energy imparted to an electron accelerated across a potential difference of 1 volt:

$$1 \text{ eV} = 1.6 \times 10^{-19} \text{ J} \quad \text{------ (1)}$$
$$= 1.6 \times 10^{-12} \text{ erg}$$
$$= 4.4 \times 10^{-26} \text{ kWh}$$

The electron volt describes potential as well as kinetic energy. The binding energy of an electron in an atom is a form of potential energy. Kiloelectron volt (keV) and million electron volt (MeV) are also used as units in practice; 1 keV = 10^3 eV and 1 MeV = 10^6 eV.

Neutrons

Neutrons were discovered by James Chadwick in the year 1932. When beryllium ($^9_4$Be) was bombarded by alpha (α) particles ($_2^4$He), a highly penetrating radiation ($_0^1$n) was emitted with release of energy (Q_4). Chadwick concluded that these particles possess mass nearly equal to proton and have no charge. He called them neutrons and the reaction can be written as:

$$_4^9\text{Be} + _2^4\text{He} \longrightarrow _6^{13}\text{C} \longrightarrow _6^{12}\text{C} + _0^1\text{n} + Q_4$$
------ (2)

Properties of Neutrons

1. Neutrons are neutral particles with zero charge and mass slightly greater than proton.
2. Neutrons are not affected by electric and magnetic fields.
3. Neutrons are the constituents of all nuclei except $_1^1$H.
4. Neutrons are stable inside the nucleus, but outside they exist for a short time and the half-life of a neutron is 13 minutes.
5. As neutrons are uncharged particles, they can easily penetrate into other nuclei.
6. In heavier nuclei, the number of neutrons is more than that of protons.
7. Neutrons with energies between 0 and 1,000 eV are called slow neutrons. Neutrons, whose energies match those of the surrounding atoms, are known as thermal neutrons. Neutrons with energies 0.5–10 MeV are called fast neutrons.

ELECTROMAGNETIC RADIATION

An electric charge is surrounded by an electric field. If the charge moves, a magnetic field is produced. When the charge undergoes acceleration or deceleration, the magnetic and electric fields of charge will vary. The combined variation of the electric and magnetic fields result in loss of energy. The charge radiates this energy in a form known as electromagnetic radiation. The electromagnetic radiation moves in the form of sinusoidal waves (Fig. 2.2). The nature of

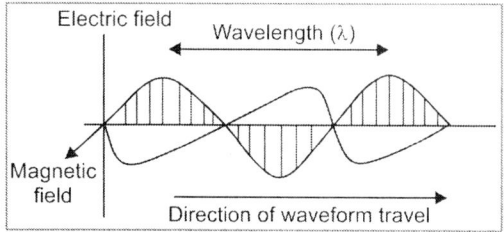

FIGURE 2.2: Electromagnetic wave

the electromagnetic radiation (X-rays, ultraviolet, etc.) depends on the way in which the electric charges are disturbed. Electromagnetic radiations are transverse waves, which transfer energy away from the electric charge. Electromagnetic radiations may be absorbed or scattered in a medium, resulting in loss of energy.

Wave Characteristics

Electromagnetic waves possess wavelength (λ), frequency (ν) and velocity (c). The distance between two consecutive positive peaks is known as wavelength. The number of cycles of the wave, which pass a fixed point per second, is known as frequency of the wave. The velocity of the wave is the distance traveled per second by the wave. The relation between wavelength, frequency and velocity of the electromagnetic wave can be expressed as:

$$c = \nu\lambda$$
------ (3)

All electromagnetic waves, travel at the same velocity in a given medium. In vacuum, the velocity is about 2.998×10^8 m/s. The wavelength of X-rays and gamma (γ)-rays are in the order of nanometer (nm).

Particle Characteristics

Though electromagnetic radiations have the properties of waves, they also behave like a stream of small bullets (particle), each carrying a certain amount of energy. This bundle of radiation energy is called quantum or photon. The amount of energy carried by the photon depends upon the frequency of the radiation.

The actual amount of energy (E) carried by a photon is given by the equation:

E = hν

where,

h is the Planck's constant = 6.63×10^{-34} Js.

Substituting the value of ν = c/λ in the above equation, then the energy:

E(keV) = hc/λ = 1.24/λ ------ (4)

where,

λ is in nanometer (nm).

The product of the velocity of light (c) and Planck's constant (h) is 1.24. It is seen that the energy of the photon is inversely proportional to its wavelength and as the wavelength decreases, the energy increases.

Mass-energy Equivalence

Einstein's theory of relativity states that mass and energy are equivalent and inter changeable. In any reaction, the sum of the mass and energy must be conserved. Einstein showed that the speed of some nuclear processes approach the speed of light. At this speed, mass and energy are equivalent:

E = mc² ------ (5)

where,

'E' represents the energy equivalent to mass 'm' at rest and 'c' is the speed of light in a vacuum. For example, the energy equivalent of an electron of mass 9.109×10^{-31} kg is:

E = mc² = 9.109×10^{-31} kg × $(2.998 \times 10^8)^2$

= 0.511 MeV

Electromagnetic Spectrum

Electromagnetic spectrum includes radiowaves, microwaves, infrared, visible light, ultraviolet, X-rays, γ-rays and cosmic rays (Fig. 2.3). All of them travel at a velocity 'c' in a vacuum. The wavelength and photon energy of the whole range of electromagnetic radiation are summarized in Table 2.2.

Ionizing Radiation and Non-ionizing Radiation

Ionization is a process of removal of electron from neutral atom. The radiation that does ionization in a medium by removal of electron is called ionizing radiation, e.g. X-rays, and γ-rays has sufficient energy to do ionization. As a result, ionized atoms and molecules or ion-pairs are produced. This forms the basis for biological effects of radiation. Radiation that do not have sufficient energy to produce ionization are called non-ionizing radiation, e.g. visible light, infrared, radiowaves and television broadcasts, etc.

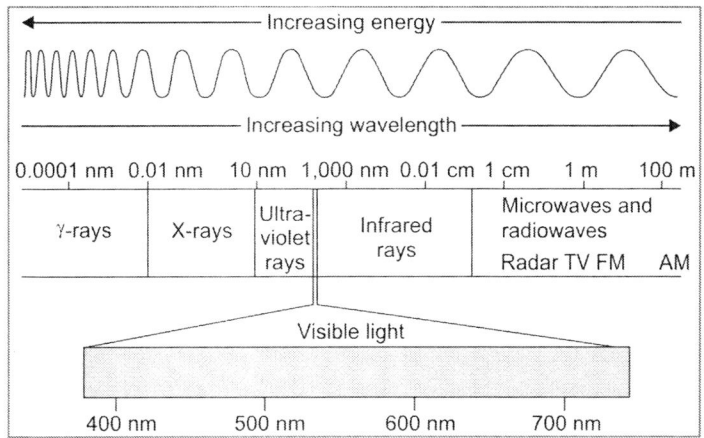

FIGURE 2.3: Electromagnetic spectrum

Table 2.2: Electromagnetic radiation and their classification

Radiation	Wavelength	Frequency	Energy
Radiowaves	1,000–0.1 m	0.3–3,000 MHz	0.001–10 µeV
Microwaves	100–1 mm	3–300 GHz	10–1,000 µeV
Infrared	100–1 µm	3–300 THz	10–1,000 meV
Visible light	700–400 nm	430–750 THz	1.8–3 eV
Ultraviolet	400–10 nm	750–30,000 THz	1.8–100 eV
X- and γ-rays	1 nm – 0.1 pm	$3 \times 10^5 - 3 \times 10^9$ THz	1 keV – 10 MeV

Fluorescence

When electromagnetic radiation falls on a phosphor, visible or ultraviolet light is emitted from the phosphor and it is called luminescence. The electromagnetic radiation raises the valence electrons to the conduction band, which return to the valence band to fill up the holes. As electron falls through the luminescence centers, they emit the surplus energy in the form of flashes of light, which is called luminescence. If the luminescence is instantaneous, within 10^{-8} s, it is called fluorescence.

The energy of light emitted depends on the difference in energy across the luminescence centers. It is always less than the energy, which originally stimulated the fluorescence, e.g. a phosphor exposed to ultraviolet may emit visible light. Fluorescent phosphors such as thallium-activated sodium iodide (NaI:Tl, gamma camera), terbium-activated gadolinium oxysulfide (intensifying screen) and sodium-activated cesium iodide (image intensifier) are used in diagnostic radiology.

If the emission of light is delayed beyond 10^{-8}s, it is called phosphorescence. When the valence electrons are stimulated, they get trapped in the conduction band. They acquire energy from the atom (internal energy) and return to the valence band by emitting luminescence. It is a random process, which takes time to accomplish. The emission of light decays exponentially with time constant that depends upon the temperature of the phosphor.

Inverse Square Law

The intensity of electromagnetic radiation is inversely proportional to the square of the distance from its source. Let us consider a point source 's', emitting radiation at constant rate. The radiation spread over the inner surface of an imaginary sphere of radius 'd' with surface area $4\pi d^2$. Then, the radiation intensity (I) at a point 'd' is given by the relation:

$$I \alpha \frac{1}{d^2} \qquad \text{------ (6)}$$

The inverse square law is based on the following assumptions:
- The source of radiation is a point source
- The radiation travels in straight line
- The radiation is emitted equally in all directions
- The energy is radiated at a constant rate
- No radiation energy is lost on its way from the source to the point of measurement.

RADIATION UNITS

Exposure (Roentgen)

The term exposure (X) refers the radiation quantity measured in terms of ionization in air, in a small volume around a point. Exposure is a source-related term. Exposure from an X-ray source obeys inverse square law. The unit of exposure is Roentgen (R).

One Roentgen shall be the quantity of X- or γ-radiation such that the associated corpuscular emission per 0.001293 g of air (1cc of dry air at NTP), produces in air, ions carrying 1 esu of quantity of electricity of either sign. The unit may also be defined in terms of SI unit as:

$1R = 2.58 \times 10^{-4}$ C/kg of air ------- (7)

There are some difficulties in the unit of Roentgen. It is not a unit of dose, which is a measure of absorbed energy. It can be used only up to photon energy of 3 MeV. It is defined only for X- and γ-radiations in air.

Kerma

Kerma (K) stands for kinetic energy released in the medium, which describe the initial interaction of the photon with an atom that takes place in the medium. When radiation interacts with matter, the uncharged particles (photons and neutrons) transfer kinetic energy to the charged particles (e and P). Kerma is the measure of kinetic energy transferred to the charged particles. It is defined as the sum of the initial kinetic energy of all the charged ionizing particles, liberated by photons in a material of unit mass. The unit for kerma is Joule per kilogram (J/kg). The SI unit is gray (Gy) and the special unit is rad (r).

Absorbed Dose (Rad/Gray)

The term absorbed dose (D) refers to the amount of energy absorbed per unit mass of the substance. The unit of absorbed dose is rad, which means radiation absorbed dose. 1 rad = 100 ergs/g. This unit is independent of type of radiation and the medium. The SI unit of absorbed dose is Gy:

1 Gy = 1 J/kg ----- (8)

The unit rad is related to gray as 1 Gy = 100 r.

RADIATION INTERACTION WITH TISSUE

When X- or γ-radiation passes through a medium, it interacts with an atom and produces moving electrons. These electrons travel in the medium, interact with other atoms and produce ionization and excitation. As a result, energy is deposited on the cells, which are either damaged partially or completely. In addition, sufficient amount of heat is also produced. In summary, the X- or γ-photons transfer energy to the electrons, which intern transfer the energy to the cell system and produce the biological effect. That is why, they are called indirectly ionizing radiations.

The above interaction is said to have wave-like and particle-like properties. X- and γ-rays interact with structures that are similar in size to their wavelength. Low-energy photons tend to interact with atoms, medium energy to that of electrons and high-energy photons with that of nuclei. The above structural level interactions may be performed by five mechanisms such as:
- Coherent scattering
- Photoelectric absorption
- Compton scattering
- Pair production
- Photonuclear disintegration.

Attenuation

When radiation passes through a medium, it is partly absorbed, scattered and partly transmitted through the medium. The term attenuation refers both absorption and scattering. The attenuation is due to interaction of radiation with the medium. As a result, the intensity of radiation decreases exponentially. This means that the first few centimeter of the medium attenuates the radiation heavily, whereas later thickness attenuates lesser. To attenuate the radiation completely, one must have infinite thickness of the medium. Hence, the intensity of transmitted radiation is always lesser than the incident radiation. If a photon passes through an absorber of thickness 'X' (Figs 2.4A and B), then the incident

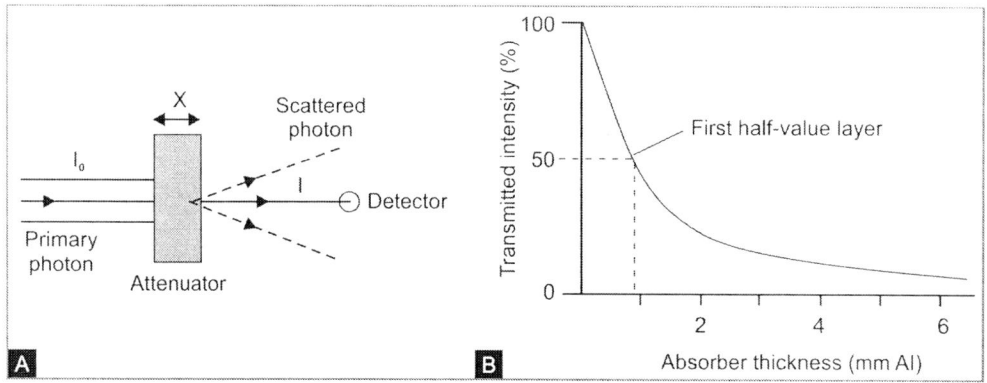

FIGURES 2.4A and B: Attenuation of radiation in a medium. **A.** Radiation interaction with an attenuator; **B.** Relationship between transmitted intensity and thickness of the absorber.

and transmitted radiation intensity is related as follows:

$$I = I_0 e^{-\mu x} \text{ or } \mu x = \log(I_0/I) \quad \text{------ (9)}$$

where,

'I' is the transmitted photon intensity, 'I_0' is the incident photon intensity, 'e' is the base of natural logarithm (2.71) and 'μ' is the linear attenuation coefficient of the medium. The negative sign indicates the reduction of radiation intensity.

Linear Attenuation Coefficient

The linear attenuation coefficient (μ) is defined as the reduction of radiation intensity per unit path length and it is expressed in cm^{-1}. This means that the 'μ' is the amount of photons removed in 1 cm travel in the medium. It depends on the medium density and energy of the radiation. Different radiation energy is attenuated differently by a same medium. Same radiation can be attenuated differently by different medium. The linear attenuation coefficient concept is very much useful to understand computed tomography (CT) scan principle and room shielding design in radiation safety.

Mass Attenuation Coefficient

Mass attenuation coefficient (μ_m) is the ratio of linear attenuation coefficient and density. It is obtained by dividing the linear attenuation coefficient by the density of the medium. The advantage is that it becomes density independent and its unit is square meter per kilogram (m^{2}kg^{-1}).

Energy Dependence

If the radiation energy is monochromatic, then its energy will remain the same, even after transmission in a medium. This means that the incident and transmitted radiation will have the same energy. The relationship between I, I_0, x and μ, described above is valid only for such conditions. If the radiation is heterogeneous, the low energy is attenuated heavily than the high energy. As a result, the effective energy of the radiation increases, which is called beam hardening. In this case, the energy of the incident and exit beam will be different. In such cases, the relationship between I, I_0, x and μ, described above may not be true.

In diagnostic X-rays, the beam is heterogeneous, but it is heavily filtered. Hence, it is assumed to be monochromatic beam and the above relationship is valid with good approximation.

Half-value Thickness and Tenth Value Thickness

In the case of heterogeneous beam, description of quality is difficult, as multiple energies are present. Hence, the term half-value thickness (HVT) is used to describe beam quality. The thickness of material that attenuates an X-ray beam by 50% is called half-value thickness or half-value layer (HVL). For a given medium, it is different for different energies. For a given energy, it is different for different medium. Since X-rays are heterogeneous, HVT is described on the basis of effective energy. The effective energy of a heterogeneous photon beam is that energy of the homogeneous beam at which the linear attenuation coefficient of the homogeneous beam is equal to the heterogeneous beam. The linear attenuation coefficient is related to the HVT as follows:

$\mu_{eff} = 0.693/HVL$ ----- (10)

Half-value thickness depends on the atomic number and density of the medium. It increases with increase of photon energy. For a given material thickness, low-energy photons undergo more attenuation than high-energy photons. For diagnostic X-ray beam energies, the HVL for soft tissue ranges from 2.5 to 3.0 cm.

The thickness of the material, which reduces the photon intensity to one tenth of its original value, is known as tenth value thickness (TVT). That means that the TVT reduces the photon intensity to 10%, whereas 90% is attenuated. However, this relationship is true only for homogeneous beam, where the attenuation is truly exponential. The TVT and HVT can be related as follows:

TVT = 3.3 × HVT ----- (8)

The concept of TVT and HVT are useful in shielding calculations in radiation safety.

Attenuation Process

The attenuation of photon beam is caused by four important processes such as:
- Coherent scattering
- Photoelectric absorption
- Compton scattering
- Pair production and photonuclear disintegration.

The Compton scatter and the photoelectric absorption are the two most important interactions in diagnostic radiology. At low photon energies, the photoelectric interaction dominates the attenuation process, whereas Compton scatter present at all energies. Pair production dominates at very high energies of radiation.

Coherent scattering

In this process, the photon interacts with the electron of an atom and sets them into vibration at the frequency of radiation. A vibrating electron, since it is a charged particle, deflects radiation in different direction. Thus, in coherent scattering, the photon undergoes a change in direction without change in wavelength (Fig. 2.5). In this process, no energy is transferred and no ionization occurs. This process occurs in low-energy radiation (15–30 keV), where the energy of the photon is small compared to the ionization energy of the atoms. The scattering angle increases with increase of photon energy. This is also

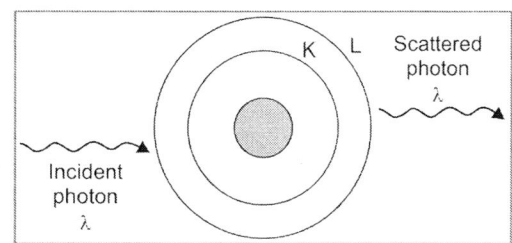

FIGURE 2.5: Coherent scattering

called unmodified scattering. Coherent scattering contributes only < 5% in diagnostic radiology and not much important in image formation. However, its influence is there in mammography.

Photoelectric absorption

In the photoelectric effect (PE), a photon of energy 'E' collides with an atom and ejects one of the bound electrons from the 'K' or 'L' shells (Fig. 2.6). The ejected electron is called photoelectron and it has kinetic energy equal to E orbital-binding energy. In this process, all the incident photon energy is transferred to the electron. The incident photon must have energy equal or greater than the orbital binding energy of the electron to perform photoelectric effect.

After the photoelectric effect, the atom is said to be ionized and there is vacancy in the orbit. This vacancy is filled by an electron of lower binding energy from higher orbit. This will create a cascade of electron transition event from outer orbit to inner orbit. The difference in binding energy is released as characteristic X-rays or Auger electrons. The photoelectric effect involves tightly bound electrons. The tightly bound electrons are mostly available in the K shell, and hence, most photoelectric interactions occur at the K shell.

The probability of photoelectric cross section per unit mass is proportional to Z^3/E^3, where Z is the atomic number and E is the incident photon energy. As the X-ray photon energy increases, the photoelectric effect decreases, resulting in decreased subject contrast. As the atomic number increases, the photoelectric absorption increases and the subject contrast also increases. That is why, barium (Z = 56) and iodine (Z = 53) are used as contrast agents, which maximize the photoelectric effect.

Even though photoelectric effect decreases with increase of energy, there are few exceptions. The relation between probability of photoelectric absorption with photon energy is plotted as exponential curve (Fig. 2.7). This means that as the photon energy increases, the attenuation decreases exponentially. However, if the photon energy is equal or greater than the K shell-binding energy, the absorption increases steeply and the curve appear with discontinuities. The above sharp discontinuity in the attenuation plot is called K-edge absorption. The elements exhibit K absorption edges, e.g. the binding energy of the K shell in iodine is 33.2 keV, which will have six times higher absorption at the K edge.

The photon energy corresponding to an absorption edge increases with atomic number. The absorption edges of elements H, C, N and O present in the soft tissue are well below < 1 keV. The iodine, barium and lead absorption edges are 33.2, 37.4 and 88 keV, respectively. The photoelectric effect is very important in soft tissue imaging for photon energy < 50 keV. It will differentiate attenuation between two tissues with slightly varying atomic number.

Compton scattering

In Compton scattering, a photon interacts with a free electron (valence) of an atom and gets scattered with partial energy (Fig. 2.8). The other part of energy is given to the valence electron, which is ejected from the atom. The ejected electron loses energy further by ionization and excitation of atoms in the tissue, resulting to patient dose. The scattered photon may travel in the medium with or without

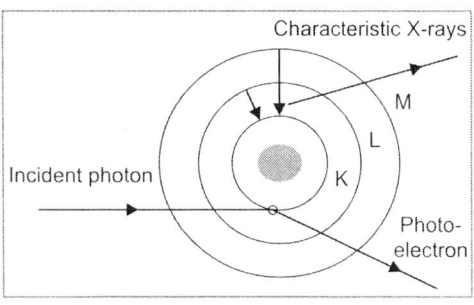

FIGURE 2.6: Photoelectric effect

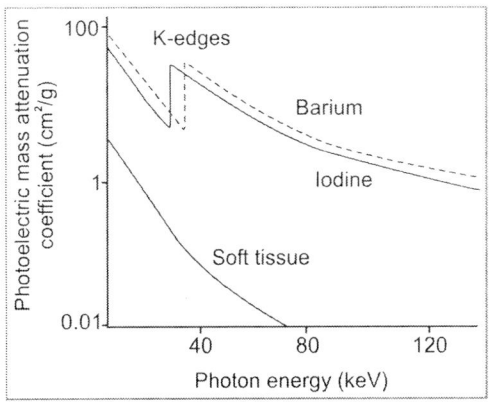

FIGURE 2.7: Photoelectric absorption as a function of photon energy

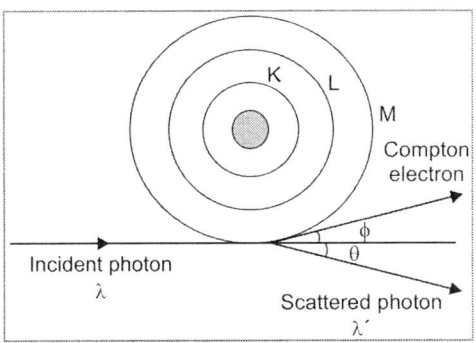

FIGURE 2.8: Compton scattering

interaction in the medium by Compton scattering or photoelectric absorption. The scattered photon will have longer wavelength compared to the incident photon. The energy of the incident photon (E_0) is equal to the sum of the energy of the scattered photon (E_{sc}) and the kinetic energy of the ejected electron:

$$E_{sc} = \frac{E_0}{1 + [E_0 (1 - \cos\theta) \div 511 \text{ keV}]} \quad \text{--- (9)}$$

where,

θ is the angle of scattered photon. As the incident photon energy increases, both photons and electrons are scattered in forward direction. The fraction of energy transferred to the scattered photon decreases with increase of incident photon energy, for a given angle of scatter.

If the photon makes a direct hit on the electron, the electron will travel straight forward (Φ = 0) and the scattered photon will be scattered back with θ = 180°. In this collision, the electron will get its maximum energy, while the scattered photon goes with minimum energy. If the photon makes a grazing hit with the electron, the electron will be emitted at right angles (Φ = 90°) and the scattered photon will go in the forward direction (θ = 0). In this collision, the electron receives minimal energy and the scattered photon goes with maximum energy.

The Compton scattering involves an interaction between a photon and a free electron resulting in ionization of atom. The incident photon energy is shared between the scattered photon and ejected electron. The probability of Compton scattering depends upon the electron density (number of electrons per gram × density) in the medium. Except hydrogen, the electron density is constant for tissue, hence it is independent of atomic number (Z). Probability of Compton interaction decreases with increase of X-ray energy and it is ∝ 1/E. The probability per unit volume is proportional to the density of the material. Hydrogenous material has higher probability for Compton scattering.

Compton scattering occurs in all energies in tissue and it is important in X-ray imaging. It is a predominant interaction in the diagnostic energy range with soft tissue (100 keV–10 MeV). Scattered X-rays provide no useful information, only reduce image contrast, create radiation hazards in radiography and fluoroscopy. In fluoroscopy, large amount of radiation is scattered from the patient and contribute to occupational radiation exposure.

Pair production

A photon having energy > 1.02 MeV, which passes near the nucleus of an atom is subjected to strong nuclear field (Figs 2.9A and B). The photon may suddenly disappear and create a positron and electron pair, which is

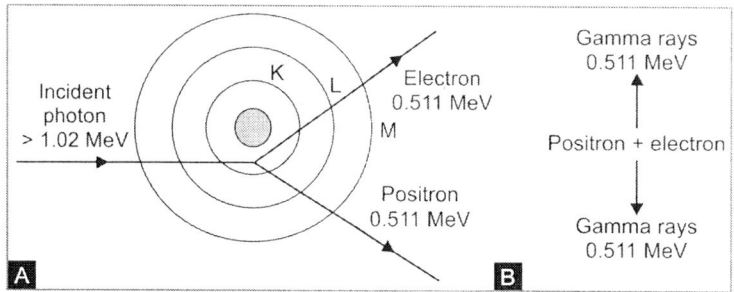

FIGURES 2.9A and B: Pair production and annihilation events. A. Pair production; B. Positron annihilation.

called pair production. For each particle, 0.511 MeV energy is required and the excess energy > 1.02 MeV would be shared between the positron and electron as kinetic energy. Actually, the interaction is between a photon and the nuclear field. This process is an example for the conversion of energy into mass as predicted by Einstein. The threshold energy for the pair production process is 1.02 MeV. The probability of pair production increases with energy for a given material and also increases with atomic number (Z^2). It is very important for photons having energy > 5 MeV.

The electron loses energy in the medium by further excitation and ionization, and filling vacancy in the orbital shells. The positron travels in the medium and loses its energy by ionization, excitation and Bremsstrahlung process. Finally, the positron combines with a free electron and produces two photons of energy of 0.511 MeV, which are ejected in opposite directions. The above process is called positron annihilation. This is an example of conversion of mass into energy and is the basis for positron emission tomography.

Relative importance of attenuation process

The total attenuation coefficient in tissue for a given energy is caused by photoelectric absorption coefficient (τ), Compton attenuation coefficient (σ) and pair production coefficient (π). The total linear attenuation coefficient in tissue (μ_{tot}) is given the relation:

$$\mu_{tot} = \mu_\tau + \mu_\sigma + \mu_\pi \ \text{------ (10)}$$

At low photon energy (< 26 keV), photoelectric process dominates the attenuation in soft tissue. As the energy increases, photoelectric absorption decreases. It is known that photoelectric absorption vary with Z^3. Hence, photoelectric absorption is more important, than Compton scattering, in high Z materials and in low-energy X-ray photons. It is the main mode of interaction with bone, contrast materials and screen phosphors in diagnostic radiology.

Compton process is present at all energies and it is very important than photoelectric absorption with low Z materials and in high-energy X-ray photons. Both the processes are equally important at:
- 30 keV energy for air, water and soft tissue
- 50 keV energy for bone
- 300 keV energy for iodine and barium contrasts.

The pair production is not much important in diagnostic radiology, but is important in mega voltage radiotherapy.

Overall in the diagnostic X-ray energy, Compton dominates in air, water and soft tissue. Photoelectric absorption dominates in contrast media, lead and screen-film systems. However, both processes are very important in bone in imaging.

In mammography, about 75% of the attenuation is caused by photoelectric absorption and the rest is due to Compton scatter. However, in chest radiography and nuclear imaging 15-20% is caused by photoelectric and the rest is due to Compton scatter. In radiotherapy, the beam energies are > 1 MeV, hence, Compton and pair production dominate the attenuation process.

Importance of Interaction in Tissue

Differential Absorption

When X-ray passes through the human body, it partly interacts with Compton scattering, photoelectric effect and partly transmitted through the body without interaction.

The Compton scattered X-ray do not give useful information for image formation. But, it creates noise by degrading the diagnostic image. Hence, suitable techniques are used to reduce scatter radiations that reaches the detector.

Photoelectric effect gives diagnostic information and helps the detector for image formation. Bone-like anatomic structures are radio-opaque and show high absorption characteristics, resulting in light areas (white) in the radiographs. The X-rays that are transmitted through the body without interaction, reaches the detector, resulting in dark areas (black) in the radiograph. The anatomical structures appear radiolucent to above X-rays.

Hence, the radiographic image is due to the difference between the X-rays that are absorbed by photoelectric process and those transmitted without interaction. This difference is called differential absorption. Reduction of kVp increases differential absorption and image contrast, with high patient dose.

Atomic Number

The probability of photoelectric absorption is $\propto Z^3$ of the soft tissue. The atomic number of bone and soft tissues are 13.8 and 7.4, respectively. The probability of photoelectric effect in bone is seven times $(13.8/7.4)^3$ more than in soft tissue. This probability decreases with increase of energy. Hence, in high X-ray energy, only few interactions occur and more X-rays are transmitted without interaction.

The Compton effect is independent of Z of the tissue. The probability of Compton scatter is equal in bone and soft tissue, and decreases with increasing energy. This decrease is slow compared to photoelectric absorption, which decreases rapidly. Hence, Compton scatter dominates at high photon energies.

Mass Density

The interaction of X-rays with tissue is proportional to the mass density, regardless of the type of interaction. Mass density is the quantity of matter per unit volume expressed in kg/m^3. It is related to mass of each atom and explains how tightly the atoms are packed. The atomic number and mass density of various tissues are given in Table 2.3. If the mass density increases, the electron number increases, which accounts for higher interaction.

Table 2.3: Atomic numbers and mass density of various tissues and contrast agents

Substance	Effective atomic number (Z)	Mass density (kg/m³)
Lung	7.4	320
Fat	6.3	910
Soft tissue, muscle	7.4	1,000
Bone	13.8	1,850
Air	7.6	1.3
Barium	56	3,500
Iodine	53	4,930

In addition to the Z-related photoelectric effect, mass density also contributes to differential absorption. The X-rays are absorbed and scattered two times (1,850/1,000 = 1.85) in bone than in soft tissue. Mass density helps in imaging lungs in radiography. In case of air (Z = 7.6) and soft tissue (Z = 7.4), the atomic numbers are almost same. But an air-filled soft tissue cavity can be imaged, due to their mass density difference.

The contrast agents barium and iodine has high atomic number and high mass density. Hence, they can be used with low kVp technique to see internal organs. However, use of high kVp technique will help in visualizing the lumen of the organ. In this case, the X-ray penetrates the contrast and helps to outline the organ.

Photon Energy

In diagnostic radiology, the kilovoltage range is about 20–150 kVp and effective photon energy ranges from 15 to 100 keV. The relative importance of interaction changes over this range. At low energy, photoelectric absorption is the main cause of attenuation than Compton scatter. Hence, soft tissue and bone appear as dark and light in the X-ray film. The thickness of tissue is also important in the attenuation process. Thick layer of soft tissue and thin layer of bone may produce same amount of attenuation. However, the use of low energy with photoelectric effect interaction can differentiate the above, overcoming the thickness effect.

In high energy, Compton scatter is dominant and the differentiation between soft tissue and bone reduces. Though Compton scatter depends upon density, tissue differentiation is still possible with use of high kVp techniques.

PARTICLE INTERACTIONS

Particle radiation includes α- and β-particles, protons, electron, positron and neutrons. In general, particles are classified as heavy charged particle, charged particle and uncharged particles. Alpha and protons are heavier particles compared to that of electrons and positrons. The behavior of heavy particles is different from lighter particles. Electron is a lighter charged particle and neutron is an uncharged particle.

Charged particles interact with matter with electrical forces and lose energy in a medium by ionization and excitation, i.e. the charged particles collide with orbital electron and lose energy. This means that the coulomb force between the electric fields of the charged particle and the atomic electrons perform the interaction. In excitation, the charged particle transfers its energy to the orbital electron and moves it to the higher orbits. The transferred energy should be less than the binding energy of the electron. The excited electron, return to lower energy by emission of characteristic X-rays and Auger electrons. About 70% of the charged particle energy is spend via excitation.

If the energy of the particle exceeds the binding energy of the electron, ionization occurs, where an electron is ejected from the atom. In ionization, an electron and a positive atom are produced, which are called ion pairs. Sometimes, the ejected electron produces further ionization known as secondary ionization. These electrons produced by secondary ionization are called delta (δ) rays. The energy required to produce an ion pair in soft tissue is 34 eV.

The ionization and excitation occur along the path of a charged particle in a medium and the rate at which the ion pairs (IP) are formed depends upon the charge and energy of the incident particle and the atomic number of the medium. In addition, collision between the charged particle and atomic nuclei may also take place, resulting in Bremsstrahlung X-rays. This is called radiative loss of energy, more likely in lighter charged particle such as electrons. In addition, electrons undergo multiple scattering in the medium.

Specific Ionization and Bragg Peak

The number of primary and secondary ion pairs produced per unit path length of charged particle is called specific ionization, expressed in IP/cm. Specific ionization increases with particle charge and decreases with particle velocity. Larger particle interacts with greater coulombic field, lose energy and slows down. It permits greater time for interaction. Thus, the specific ionization of α-particle is higher than that of proton and it is in the order of 7,000 IP/mm. The energy loss in the medium can be also specified by linear stopping power (S). It is the rate of kinetic energy loss in the medium per unit path length. This can be modified as mass stopping power (S/ρ) and expressed in MeV cm^{-2} g^{-1}.

The plot of specific ionization as a function of particle range in the medium is shown in the Figure 2.10. The particle enters the medium with a velocity and causes some specific ionization, leading to energy loss. The rate of energy loss is proportional to the particle charge and inversely proportional to its velocity. As it travels in the medium, its velocity decreases, spends much time in the medium and increases its specific ionization. When the velocity is zero, the specific ionization increases to a maximum called Bragg peak. Beyond the peak, the particle picks up electrons and become electrically neutral; hence the specific ionization decreases rapidly. This property of the particle is used to treat cancer patient in radiotherapy. The particle delivers maximum radiation at a given depth without exit dose. Protons are the well-established external beam particle recommended for precision radiotherapy. The Bragg peak can be modulated and flattened by using a filter to suit the clinical conditions.

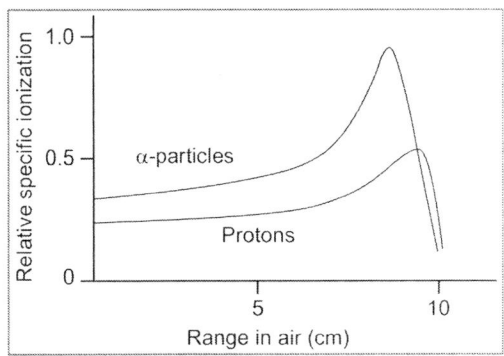

FIGURE 2.10: Bragg curves for protons and α-particles

Electron Interaction

Electrons, on the other hand, has small mass and undergo multiple scattering in their track. They do not exhibit Bragg peak and mainly deposit energy by ionization and excitation. The types of electron interactions include:
- Inelastic collisions
- Elastic collisions.

Inelastic Collision

Here, part of the kinetic energy of the electron is lost during ionization and excitation, resulting in Bremsstrahlung or secondary electrons or delta rays. Inelastic collision with atomic nuclei causes Bremsstrahlung production, whereas inelastic collision with orbital electrons causes secondary electrons or delta rays. The former effect is dominating and the latter is very rare event.

Elastic Collision

Here, there is loss of energy and the colliding particle gets deflected to a different direction, leading to scattering. This is also possible with atomic nuclei called coulomb scattering as well as with orbital electrons called electron-electron scattering. Elastic collision may be multiple, resulting in multiple small angle deflections. In this type, the electron undergoes zig-zag path and lose energy by inelastic collisions. The overall mass stopping power of electron (Tot) consists of both collisional (col) and Bremsstrahlung (rad) components and usually written as given below:

$$(S/\rho)_{Tot} = (S/\rho)_{col} + (S/\rho)_{rad} \quad \text{----- (11)}$$

The linear stopping power depends on electron density (electrons per cubic centimeter), whereas the mass stopping power vary with electrons per gram. Since, the low atomic material have high electrons per gram, their mass stopping power is high (e.g. gas). The total stopping power of electron varies with energy, but electron energy is > 1 MeV and is almost constant in water. The radiative stopping power is proportional to energy and Z^2. At higher energy, the radiative stopping power is greater with high Z materials.

Multiple scattering and energy degradation make the electron beam more heterogeneous in energy with angular spread. This depends on the density and composition of the medium in which the electron travels. To explain this behavior, mass angular scattering power is defined, which is the mean square scattering angle per density per unit path length, in analogy to mass stopping power.

Proton Interaction

Proton is a heavy and positively charged particle with charge and mass of 1.6×10^{-19} coulombs and 1.6×10^{-27} kg respectively, with a half-life of 10^{32} years. It interacts mainly with:
- Inelastic collision
- Elastic collision
- Nuclear reaction.

The inelastic collision happens with orbital electrons and atomic nuclei. The former interaction is dominant one and performed under ionization and excitation, whereas the latter process is less significant.

The stopping power of proton is related to its LET, which is the energy transferred to the medium per unit path length. The LET do not account the energy loss by the delta rays. LET of a charged particle is proportional to its charge square and inversely proportional to its kinetic energy. It is the product of specific ionization and the average energy deposited per ion pair. The LET explains the energy deposition density and largely determines the radiobiological effect of radiation. Thus, the LET of the particle depends on its RBE value and the RBE of proton is 1.1.

Mass stopping power is proportional to its square of the velocity. As the velocity increases, the energy loss is high. When the velocity becomes zero, the energy loss is maximum. Mass stopping power is greater in low Z material compared to high Z, since proton undergoes large angle scatter in high Z material. Thus, low Z material decreases proton energy with minimal scatter, whereas high Z material offers more scattering with minimum loss of energy. Proton Bragg peak usually follow sharp peak with narrow width, not suitable for clinical treatment. Hence, the Bragg peak is spread out by superimposing beams of different monoenergetic proton energies. This will make the Bragg peak wider and usually called spread out Bragg peak (SOBP), which is more suitable for clinical treatments. Usually, high-energy proton is placed at the distal end and other proton beams are added in decreasing energy order. After the SOBP, the dose becomes zero.

Neutron Interaction

Neutrons interact in three ways such as:
- Elastic collision with nuclei
- Inelastic collision with nuclei
- Neutron capture.

Elastic Collision

Here, the kinetic energy of the neutron is redistributed between neutron and the nuclei, after the collision. However, the kinetic energy of the colliding bodies is the same before and after collision. In a head on collision, the energy transfer is maximum. The energy transfer is more efficient for low Z and maximum for hydrogen, compared to high Z. Hence, shielding materials for neutron are water, paraffin wax, polyethylene and plastic, which are basically hydrogenous materials. Though lead is good shield for X-rays, it is poor towards neutron.

Inelastic Collision

Here, the neutron is absorbed by the medium with formation of compound nucleus, which is in exited state and later release neutron and γ-rays. This type of interaction mostly happens in high-energy neutron with heavier nuclei.

Neutron Capture

Neutron Capture is a similar process, but the compound nucleus emits γ-rays, e.g. Co-60 production. The probability for neutron capture is high for slow neutrons than the fast neutrons.

BIBLIOGRAPHY

1. Stewart Carlyle Bushong. Radiologic Science for Technologists: Physics, Biology, and Protection, 9th edition. St. Louis. Missouri: Elsevier Mosby; 2008.
2. Thayalan K. Basic Radiological Physics. New Delhi: Jaypee Brothers Medical Publishers (P) Ltd; 2001.
3. Thayalan K. Textbook of Radiological Safety. New Delhi: Jaypee Brothers Medical Publishers (P) Ltd; 2010.
4. Thayalan K. The Physics of Radiology and Imaging. New Delhi: Jaypee Brothers Medical Publishers (P) Ltd; 2014.

Interaction of Radiation with Cell

CHAPTER 3

SEQUENCE OF RADIATION EVENTS

Radiation can damage cell and result in biological effects in a multicellular organism. Therefore, the study of the radiation effects at cellular level is more useful to know about the radiation damage. The time interval between the radiation exposure and manifestation of biological effects can be divided into three stages (Fig. 3.1), namely:
- Physical stage
- Chemical stage
- Biological stage.

Physical Stage

The physical stage refers to the interaction between charged particles and the atoms/molecules of tissues. In this stage, the atoms or biomolecules absorb the radiation energy and undergo ionization and excitation, and release electrons. These electrons are called secondary electrons and they transfer energy to the surrounding by further excitation, ionization and thermal heating in the medium. Its duration is about 10^{-7} s and the deposition of energy is rapid and random.

The secondary electrons may produce additional low-energy electrons, known as delta (δ)-rays. About 1,000 δ-rays are produced by a single 30 keV electron. These δ-rays may cause additional excitation and ionization in the medium. Thus, very low-energy electrons are produced, whose energy is < 7.4 eV, the first excitation potential

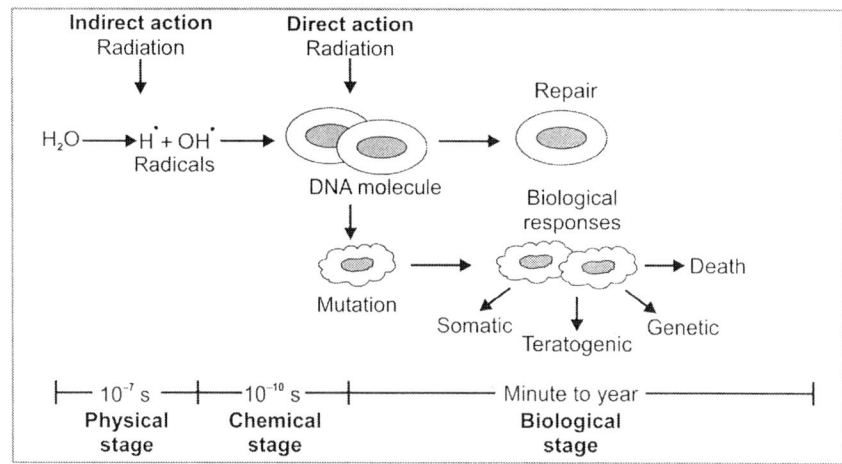

FIGURE 3.1: Interaction of radiation in cell; physical and biological response of ionizing radiation

of liquid water. They transfer their kinetic energy to water molecule by vibrational, rotational and collisional energy exchanges.

The low-energy electrons of δ-rays form a unique ionization pattern in their track over a short range of 4–12 nm. The energy deposition of about 100 eV in the shorter track is called spurs, whose diameter is about 4–5 nm. The energy deposition of about 300–500 eV in a pear-shaped track (12 nm) is called blobs. It is believed that there are spurs and blobs along the tracks of the charged particle. A spur may contain 3 ion pairs, whereas a blob may contain 12 ion pairs and deposit energy non-uniformly in the medium.

Since the dimensions of spur and blobs are similar to that of a deoxyribonucleic acid (DNA) (2 nm), there will be an overlap between them. This initiates active and multiple radical interactions with DNA strands, resulting in variety of damages. A spur can produce damage in multiple locations in the DNA, which are close to each other. The types of damages are clustered damage, complex damage and locally multiple damaged sites. In the case of low-linear energy transfer (LET) such as X- and gamma (γ)-rays, about 95% of the energy deposition is due to spurs. These spurs and blobs produce free radicals that increase the probability of damage.

Chemical Stage

In the chemical stage, the damaged atoms and molecules further reacts with the cellular components through chemical reactions. The exposed biomolecules rearrange themselves, which results in formation of primary lesions in them. Primary lesions are transformed into bioradicals, resulting in molecular alterations. Bioradicals can also be formed with the interaction of radicals with biomolecules. The structural changes include:
- Hydrogen bond breakage
- Molecular degradation
- Inter- and intra-molecular crosslinking.

The duration of the above event is about 10^{-10} s.

Biological Stage

The biological stage starts with enzymatic reactions, which act on the residual chemical damage. Majority of the lesions are repaired and lesions that are not repaired properly result in cellular death or mutations in cells. Increased number of cell death may lead to organ death that appears as clinical changes. The biological effect includes lethality, mitotic inhibition, division delay, chromosome aberrations and induction of mutations. These effects can create radiation sickness, resulting in delayed somatic effects in humans.

Cell may take time to die and may undergo mitotic division before death. The death of stem cells appears as early effects. After the cell killing, compensatory cell proliferation occurs in the tissue. Mutation can occur in both:
- Germ cells
- Somatic cells.

The mutation in germ cells may result in hereditary effects, whereas cancer induction is the outcome in somatic cells. The time schedule for biological effects is few minutes to several years. Presence of chemicals such as radioprotectors and sensitizers during the radiation exposure may alter the above response. Protectors mostly reduce indirect action by scavenging radicals, whereas sensitizers enhance the damage of the biomolecules. At the biological stage, the altered DNA molecule may be repaired by cellular repair process. This may influence post repair processes, which can modify the magnitude of the effects.

DIRECT AND INDIRECT ACTION

Direct Action

Biological effects resulting from the direct interaction of radiation with the target sites is called the direct action. There are critical sites or targets within the cells, which must be damaged in order to kill the cell. DNA is considered as critical target for radiation exposure. In a

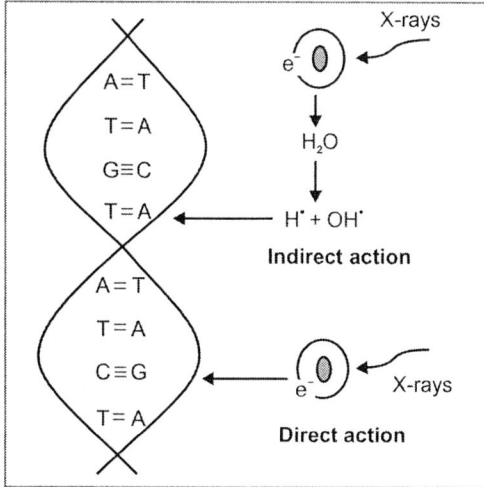

FIGURE 3.2: Direct and indirect action of radiation on a deoxyribonucleic acid (DNA) molecule

radiation interacts with oxygen and water molecules present in the cell. These interactions produce a large number of free radicals, which are atoms or molecules, or ions with an unpaired electron in the outermost shell, and hence, are highly reactive. Since electrons are spinning, a pair of electron spins both clockwise and anticlockwise, giving stability. Unpaired electron spins in one direction and is unstable. Free radical has high degree of chemical reactivity and interacts strongly with biomolecules. This is a major source of radiation damage. Human body tissue is composed of 80% water and the major interaction is indirect action (66%) through water. The effects of X- and γ-rays in macromolecules of living system are mainly due to indirect interactions.

RADIOLYSIS OF WATER

Absorption of radiation by water molecule (radiolysis) leads to ionization and results in ion pairs (H_2O^+) and electrons (e^-). The electron will leave the parent molecule and get trapped among other water molecules. The water molecule due its polar nature, tend to have its positively charged hydrogen atom facing the electron and the negatively charged oxygen atom away from the electron (Figs 3.3A to C). The electron, which is surrounded by a sphere of polar water molecules, is said to be hydrated, and hence, it is called aqueous or hydrated electron (e^-_{aq}). The e^-_{aq} is a transient species with a life time of about 1 μs in pure water and absorption band of 720 nm. It reacts with another water molecule to form negative water ion (H_2O^-) and attains stability with a life time of few microseconds.

direct action, the radiation interacts directly with the critical target and creates biological effects (Fig. 3.2). It is a dominant process with high-LET radiations such as alpha (α), neutrons, etc. It involves rupture of cell membrane and breaking of chromosome structure, resulting in DNA strand breaks. The fragments of chromosomes produced in a direct interaction, which can join together in a wrong manner, result in chromosomes with abnormal structures. This is known as chromosomal aberration. The frequency of chromosomal aberrations increases with radiation dose, and hence, the magnitude of aberrations is a biological indicator of radiation dose absorbed in human body. Chromosomal aberration analysis (CAA) is useful in determining the radiation dose received by a person who is accidentally exposed to high radiation dose (> 100 mGy).

Indirect Action

Alternatively, the radiation can interact with other molecules (e.g. water) and forms free radicals. The interaction of free radicals with critical sites may also result in biological effects, which is called indirect action. For example,

Ionization

$$H_2O \longrightarrow H_2O^+ + e^- \longrightarrow H_2O^+ + e^-_{aq} \quad ----(1)$$

$$H_2O + e^-_{aq} \longrightarrow H_2O^- \quad ----(2)$$

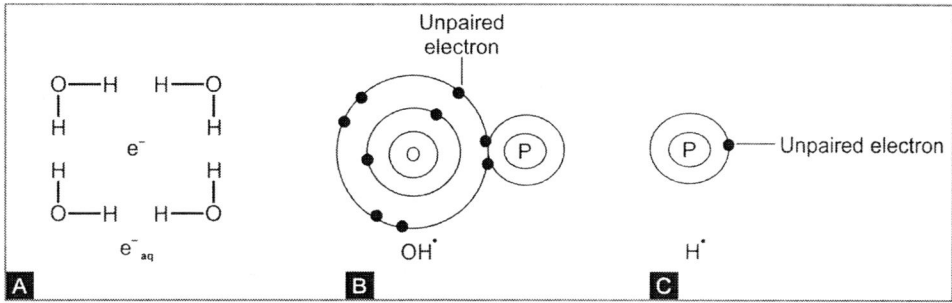

FIGURES 3.3A to C: Radiolysis of water and reactive species.
A. Aqueous electron; B. Hydroxyl radical; C. Hydrogen radical.

The H_2O^+ and H_2O^- are said to be radical ions, due to their charge and unpaired electron. The H_2O^+ is extremely reactive chemical species and performs variety of reactions. It acts as a strong oxidizing or reducing agent by combining with macromolecules. Its life time is too short, of the order of 10^{-10} s. The H_2O^+ reacts with another water molecule and produce hydroxyl radical, $OH°$. The H_2O^+ itself split into H^+ and $OH°$:

$$H_2O^+ + H_2O \longrightarrow H_3O^+ + OH°$$
(hydroxyl radical) ----- (3)

$$H_2O^+ \longrightarrow H^+ + OH° \text{ (hydroxyl radical)}$$
----- (4)

The $OH°$ has nine electrons and one electron short of the number required to give stability. It is highly reactive and has a life time of 10^{-9} s and its absorption band is about 260 nm. It can diffuse to a distance equal to twice the diameter of DNA double helix. Similarly, H_2O^- ion radical is also unstable and rapidly forms ions and H-free radicals:

$$H_2O^- \longrightarrow H° \text{ (hydrogen radical)} + OH^-$$
----- (5)

Excitation

Radiation exposure may also excite water molecules (H_2O^*), which in turn get dissociated and produces hydrogen ($H°$) and hydroxyl radicals ($OH°$), in a limited proportion:

$$H_2O \longrightarrow H_2O^* \longrightarrow H° + HO°$$
----- (6)

Thus, ion radicals (H_2O^+ and H_2O^-) and free radicals ($H°$ and $OH°$) are produced by radiolysis of water. Among this, e^-_{aq}, $OH°$ and $H°$ are the important reactive species and their relative yield are about 45%, 45% and 10%, if observed immediately after the event. These are rapidly reacting species, which can carry out variety of chemical reactions. Free radical can diffuse up to 4 nm in cells and can cause damage at distant sites. The life time of $H°$ and $OH°$ is about 10^{-10} s only and they have tendency to recombine with water rather than causing biological damage, which will be discussed in the following paragraphs.

Free Radical Reaction

The radicals can recombine with other free radicals and forms non-reactive chemical species such as H_2O and H_2O_2:

$$H° + OH° \longrightarrow H_2O \text{ (water)} \quad ----- (7)$$

$$OH° + OH° \longrightarrow H_2O_2 \text{ (hydrogen peroxide)}$$
----- (8)

The formation of water does not make any biological damage, whereas hydrogen peroxide is highly toxic to organisms. The $H°$ radical is a powerful reducing agent, whereas $OH°$ and $HO_2°$ are powerful oxidizing agents and easily combine with macromolecules.

They can attack biomolecules (DNA) via hydrogen abstraction, OH radical addition and cause damage.

The presence of oxygen enhances free radical damage. The oxygen combines with hydrogen radical to form hydroperoxyl radical ($HO_2°$):

$$H° + O_2 \longrightarrow HO_2° \text{ (hydroperoxyl radical)} \quad \text{-----(9)}$$

In the case of low-LET radiation, free radicals are the primary agents that cause the biologic effects. Approximately, two third of all radiation-induced damage is considered to be caused by the hydroxyl free radical in low-LET radiations. However, in the case of high-LET radiations the direct action dominates over the indirect action of radiation.

Repair of radiation damage is possible at molecular, cellular and tissue levels. Enzymatic mechanisms and cofactors do exists within cells, which repair the radiation damage and return to pre-irradiated state within hours to days. Repair of biomolecules can occur by hydrogen donation from thiol compounds. This is followed by the production of thiol radical, which is less damaging. For example, DNA single strand break and base damage can be repaired by specific endonucleases and exonucleases that are present in the cell. If the damage is severe, the DNA molecule may undergo mutation. However, it depends on the DNA location and its environment, radiation dose, dose rate, special distribution of energy and radiosensitivity of the system.

Cells are more radiosensitive during the phase of cell division known as mitosis, when the chromatin material is being distributed to daughter cells. Hence, rapidly dividing cells such as intestinal epithelium, bone marrow cells and reproductive cells are more radiosensitive. Highly differentiated tissues, such as muscle and brain, are least radiosensitive to radiation.

CHARACTERISTICS OF ACTIONS OF RADIATION

Direct and indirect interactions of radiation with cell ultimately result in:
1. Cell modifications such as gene mutation, cell transformation and chromosome aberrations.
2. Cell death.

Mutation in germ cell may lead to hereditary effects that carry over to many generations. Oncogenic transformation of the cell may be delayed even up to 40 years. Cell death has time range and it is expressed in hours to days.

Indirect Action

The effect of radiation by indirect action is characterized with:
- Dilution effect
- Effect of freezing
- Presence of second solute
- Solute pH.

The effect of ionizing radiation on aqueous solution depends upon the concentration of the solute. The number of molecules transformed per 100 eV of energy absorption is called G value. At low solute concentration, the radical concentration is more than the solute concentration. Radicals interact with each other and lose their number. As the solute concentration increases, the radical interacts mostly with solute and the G value increases. Thus, the G value increases with solute concentration. At a particular solute concentration, it becomes saturated. This is known as dilution effect.

The radicals have to be mobile in order to migrate and react with solute molecules. By freezing the solution, the mobility of the radicals is reduced, resulting in reduced irradiation effect. When more than one solute is present, the reactions of the one solute will be affected by the presence of radicals. Hence, there will be decrease in yield due to radical reactions. If the pH decreases more and more, e^-_{aq} is transformed into $H°$ radicals, this may cause increased inactivation especially in proteins.

Direct Action

Ionizing radiation can interact directly with biological molecules by causing ionization and excitation. One or more chemical bonds may be broken and leave the atoms or molecules with unpaired electrons. This process of radical formation takes about picosecond time. Radical-radical interaction may result in crosslinking. Radical can also react with oxygen and produce chain reactions. Further reaction with DNA, lipid and protein molecules may produce crosslinking. About one third of the radiation interaction is only by direct action. This process is dominant with high-LET radiation.

IRRADIATION OF MACROMOLECULES

The irradiation of macromolecules can produce three effects namely (Fig. 3.4):
- Main chain scission
- Crosslinking
- Point lesion.

The molecules are more radiosensitive, if irradiated in vivo. Macromolecule synthesis is more important for cell survival and reproduction. In protein synthesis, messenger ribonucleic acid (mRNA) is transcribed from DNA and it is transferred to cytoplasm. The genetic code from mRNA is then translated into protein with the help of transfer RNA (tRNA). In the case of DNA synthesis, the deoxyribose, phosphate and base molecules accumulate in the nucleus, during G1 phase. They form a single large molecule. During the S phase, the parent DNA is duplicated into two daughter DNA molecules. Thus, synthesis is carried out with series of events, irradiation of macromolecules may disturb any one of the event, in turn, the synthesis.

Main-chain Scission

Main-chain scission is the breakage of the backbone of the long-chain macromolecule. This will result in production of multiple numbers of small macromolecules. The size of the molecule is reduced and viscosity of the surrounding solution is also reduced. The measurement of viscosity of the solution may help to detect main chain scission.

Crosslinking

Some macromolecules produce small, spur-like structures in sideways after irradiation. The spur extends laterally, away from the main chain. They have stickiness at the end and are capable of attaching to a neighboring macromolecule or another segment of the same molecule. This is called crosslinking (Fig. 3.5).

Crosslinking may be protein-protein crosslink, DNA-protein crosslink, intra- and interstrand crosslink. Intrastrand crosslinking is one in which crosslinking occurs between bases on the same stand, e.g. formation of adjacent thymine dimers (TT). Dimerization reduces the distance between the bases and results in distortion (kink) of sugar-phosphate backbone. The rupture of hydrogen bond may lead to irreversible changes in the secondary and tertiary structure of DNA molecule, which may compromise genetic transcription and translation.

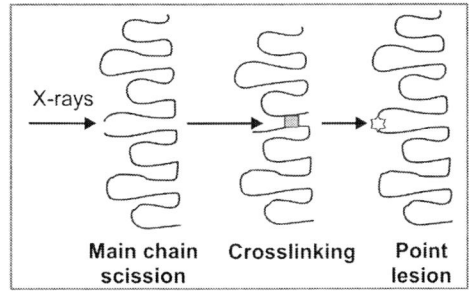

FIGURE 3.4: Radiation damage of macromolecule

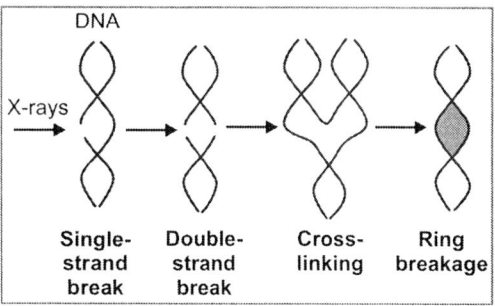

FIGURE 3.5: Radiation damage of DNA

Point Lesion

Irradiation of macromolecule may disturb a single chemical bond, known as point lesion. It is not detectable, but capable of causing slight modifications in the molecule. Hence, the molecule may malfunction within the cell.

Strand Break

Radiation exposure also causes DNA strand break in addition to base change and the DNA is degraded into smaller fragments. As a result, there will be a decrease in molecular weight of DNA. The strand breaks may be single-strand break (SSB) or double-strand break (DSB). If the break is located on one of the strand, it is referred as SSB. If two breaks are located on opposite strands, separated by five bases, then it is called DSB. The SSB between sugar and the phosphate can rejoin, if it is not separated. The SSB are mostly caused by OH radical and easily repaired compared to DSB. Presence of oxygen may cause peroxidation of base that prevents rejoining. DSB also undergo repair; if the repair is not proper then it can lead to mutation. Thus, DSB are basically genotoxic lesions that can result in chromosome alterations. This may cause activation of oncogenes, inactivation of tumor suppressor genes or loss of heterogeneity, resulting in carcinogenesis.

The DSB is the result of sparse ionization pattern of the low-LET radiation. An X-ray dose of 1 Gy, may disturb the mitotic capability in 50% of the exposed cells. All ionizing radiations are capable of producing DSB and complex DNA damages. In low-LET radiation, about one fourth of the dose is deposited in tissue via low-energy secondary electrons (0.1–5 keV). These low-energy electrons produce dense ionization tracks over a short range, equal to the diameter of double helix strand. This may result in complex DNA damages that are less likely to be repaired.

Though the amount of DSB lesions caused by the radiation is more, the number giving rise to cell killing is small. The dose of ionizing radiation that gives one lethal event per cell and leave 37% of the viable cell is called D_0. The D_0 value for low-LET X-ray lies between 1 and 2 Gy dose, which is sufficient to cause about 1,000 base damages, 1,000 SSB and 40 DSB per cell. SSB is of little importance, since its damage is mostly repaired by taking a template from the opposite strand. If incorrect repair takes place in SSB, it may lead to mutation. If both strands are broken and are well separated, it can also be repaired, since the two breaks are handled as two SSBs.

The DSB is the most important damage in DNA, after radiation exposure. The interaction of two DSB may lead to cell killing, carcinogenesis and mutation. There are many kinds of DSB, depending upon the distance between the break and kinds of end point. The DSB yield out 0.04 times that of SSB and it is induced linearly with dose. Mostly DSBs are caused by single track of ionizing radiation. DSB result in cleavage of chromatin into two pieces and is the critical lesion, responsible for cell killing. Experimental studies reveal that DSB produced initially correlate with radiosensitivity and survival at low dose. Whereas, unrepaired or misrepaired DSB correlate with survival at higher doses. High-LET radiations are capable of producing complex DSB lesions. The clusters of ionization and excitations that take place at the end of the secondary electrons tracks are capable of producing multiple lesions within 20 nm range, resulting in cell death.

The DNA damage is a normal event in a cell's life. Apart from radiation, there are many more agents that may cause DNA damage. The extent of DNA damage depends upon the repair mechanism present in the organism. Eukaryotic cells are equipped with ability to repair DSB, whereas organisms with prokaryotic cells are not equipped. A single DSB is lethal to bacteria, whereas about 60 DSB are required to kill a mammalian cell. Similarly, few SSBs are lethal to viruses and microorganisms, whereas several hundred SSB are required to kill a mammalian cell. The above difference is due to repair capacity between different organisms.

DNA Strand Break Measurements

The DNA strand break can be measured by the following methods:
- Pulsed-field gel electrophoresis
- Single-cell electrophoresis
- DNA damage-induced nuclear foci.
  1. **Pulsed-field gel electrophoresis (PFGE):** It is used to detect induction and repair of DNA DSBs. This is basically a technique of electrophoretic elution of DNA. The damaged DNA fragment is believed to have net negative charge. If it is embedded and lysed in an agarose gel, they migrate under an electric field. The electrophoresis allows the separation of DNA fragments based on their size. The fraction of DNA is directly proportional to the dose. The speed of migration is inversely proportional to their size. A DNA-specific fluorescent dye is added to study the movement. However, to improve the separation of fragments, the field is applied in alternative direction, at 60° to the axis of migration. The separation of fragments is a measure of DSB. This is calibrated with DNA of known molecular weight, e.g. yeast.
  2. **Single cell electrophoresis:** The irradiated cells are embedded in low-density agarose and lysed to release DNA. The lysis is performed under neutral buffer and alkaline buffer conditions, to quantify DSB and SSBs respectively. If the cells are not damaged, the DNA remains intact and does not migrate. If there is damage such as SSB and DSB, the DNA begins to migrate and the migration of DNA is proportional to the damage. The lysed DNA fragments take the shape of comet's tail, under electrophoresis condition. This assay has high sensitivity and specificity to SSB and a lesser degree to DSB. However, by changing the lysis condition from an alkaline to neutral pH, the sensitivity to DSB can be increased. Both methods are cell based, where the DNA is more resistive to radiation. Cell killings do not correlate with SSB, but well correlate with DSBs. This is because that DSBs are capable of causing chromosome aberrations. This technique is useful to detect DNA damage and repair, and has the advantage of finding differences in DNA damage and repair in a single cellular level. It requires only small number of cells. Hence, it has advantage for analysis of biopsy specimen, which contains small number of cells.
  3. **DNA damage-induced nuclear foci:** The irradiated cells are incubated with a specific antibody, so that they signal to a repair protein. The binding of the antibody is detected with a secondary antibody, which also carry a fluorescent tag. Under the fluorescence microscope, the location and intensity of the fluorescent tag can be quantified. The commonly assayed proteins for foci formation are γH_2AX and 53BP1. Basically, the H_2AX is a histone protein, after radiation damage, it gets phosphorylated and form γH_2AX. The 53BP1 protein also gets phosphorylated after radiation damage and form nuclear foci at the sites of DSB. Thus, this technique quantifies the DNA

damage by recruiting DNA repair proteins to the sites of DNA damage. It can be carried out on both tissue sections and an individual cell.

Mutation

Human cell contains about 30,000 genes. Each gene occupies a specific position on the chromosome. All cells have identical sets of both mother and father. An individual, therefore, receives from his/her parents, two sets of genes identical to one another. These two genes, which determine the same characteristics, are known as 'alleles'. The hereditary information contained in the gene is coded in the form of specific sequence of bases in the DNA molecule. The mutation constitutes a change in the information contained in the genes. Such changes can be brought about either by:

1. Change in the structure of the DNA molecule (base transformation, base deletion).
2. Microscopically visible changes in the structure of the chromosomes (chromosomal aberrations).
3. Changes in the number of chromosomes (aneuploidy).

Mutations of the first type are known as gene mutation or point mutation, while the latter two are called chromosome mutation. Gene mutations represent the largest category of significant genetic alterations. It can occur at random in one or both the alleles. Alleles in an unmutated state are called wild type. When one of the alleles is mutated and the other is wild type, the combination is called heterozygous. When both the alleles are of the same type, either muted or wild, it is called homozygous.

Structural and Mitotic Mutations

Chromosome mutations are involved with:
- Structural changes
- Mitotic cycle.

The first type arises from the result of breakage and reunion of the genetic material (inversions and translocations). Inversions involve breaks and rearrangements in the same chromosome, whereas in translocation, a part or full chromosome is attached to another chromosome. In the latter type, mutation occurs during the mitotic cycle and the chromosomes fail to separate from each other to form haploid germ cell. One gamete receives both chromosomes, while the other gets only one. Chromosome mutations are dominant in nature and cause both lethal- and non-lethal effects.

Thus, mutation is the term that refers the change or loss of base in a chromosome. These changes are discrete and need not have structural changes in the chromosomes. Mutations can cause heritable changes in the genetic material. Mutation can occur due to natural causes, such mutations are called spontaneous mutations. They occur at slow rate in man, which is estimated to be 1×10^5/gene/generation. The factors that influence spontaneous mutation are temperature, chemicals and natural radiation. However, radiation is capable of producing chromosome breaks, which can be viewed microscopically during mitosis.

Usually, any DNA damage can be recognized by the enzymes and gets repaired suitably. But in mutated cell, the damage is not repaired and the cell gets replicated. Though the cells have reproductive capacity, they contain altered information in their genome. The defect is passed on to the descanters of the mutated cells. However, mutations that take place in germ cells can pass the defects to the progeny of the host.

Dominant and Recessive Mutation

Mutations are classified into:
- Dominant mutation
- Recessive mutation.

In dominant mutation, a single mutant allele inherited from either parent is sufficient to cause altered phenotype. In this, the organism has one mutant and one normal allele of the gene, and is called a heterozygote. In the case of recessive mutation, two mutant alleles of the same gene, one from each parent are required to produce a mutant phenotype, and it is called homozygote. Mutations in genes that code for structural proteins are dominant, whereas the genes that code for enzymatic proteins are recessive.

Ionizing radiation can cause mutations in both germ cells and somatic cells. When radiation-induced mutation occurs in germ cells and passed on to the progeny, heritable diseases occur. The mutation in the germ cell leads to inherited effects in the offspring, known as genetic mutation. Alternatively, the mutation in the somatic cells results in effects in the individual, called somatic mutation, e.g. cancer. Radiation cannot produce new inheritable effects due to mutations, instead increases the frequency of mutations that are already occurring naturally.

EFFECTS OF RADIATION ON CELL DIVISION AND GROWTH

It is well-known that radiation results in proliferative integrity. In addition, radiation exposure can also cause delay in mitosis and cell division, which are usually termed as mitotic delay and division delay respectively. At the end of the delay period, these cells can divide indefinitely and give rise to colonies. Those which do not survive may:
1. Lyse before entering into first mitosis.
2. Divide only a limited number of times and give rise to abortive colonies containing only a few cells.
3. May continue the synthesis of cellular material in the absence of cell division, giving rise to multinucleated cells called the giant cells.

Mitotic Inhibition

Cultured mammalian cells exposed to doses above 0.5 Gy show a decrease in the percentage of cells in mitosis (mitotic index), but this change cannot be detected as in control cells. The mitotic index (MI) is defined as the ratio of the cells in mitosis to the total number of cells (N) in the sample:

$$MI = 100 \times \frac{(P + M + A + T)}{N} \quad ----(10)$$

where,

P, M, A and T are prophase, metaphase, anaphase and telophase respectively.

The cells irradiated to 0.83 and 3 Gy, show an increase in MI after a lapse of 24 hours and finally show MI higher than that of controls. Cells irradiated to a higher dose of 10 Gy do not show an increase above those of controls. The decrease in the MI seen during the first 1–2 hours is due to the blocking of the cells in G2 phase of the cell cycle. The cells in G2 phase are delayed the most by radiations as compared to cells in G1 and S phases of the cell cycle.

Further, the cells in early mitosis (prophase) revert back to G2 phase and stay there for a few hours before proceeding through division. Cells, which are in later phase of mitosis, progress without significant delay and complete division. The differential delay in different phase of the cell cycle is referred as G2 block. As soon as the cells are released from this block, they proceed through division in a synchronous fashion. This explains the increase in MI seen after 4 hours. At higher doses, however, cells in all the phases of the cycle are inhibited, and hence, the MI remains less than that of controls.

The MI is an indicator of reproductive capacity of a group of cells and is related to radiation sensitivity. Generally, cells of high MI are more sensitive to radiation. Based on the MI, cells can be divided as:

1. Zero MI, e.g. neurons, sensory organs, cornea, adrenal medulla and red blood cells.
2. Low MI, e.g. thyroid, connective tissue and vascular endothelium.
3. High MI, e.g. epidermis, bone marrow, intestinal epithelium and gonads.

Law of Bergonié and Tribondeau

The radiosensitivity of different cell line varies widely; in turn the nature of deterministic effects also varies with different organs. In 1906, Bergonié and Tribondeau (French radiobiologists) have proposed a law correlating radiosensitivity and the proliferation rate of the cells. The law states that the cells are more sensitive to radiation, when they are:
- Rapidly dividing
- Undifferentiated
- Have long mitotic future.

The law came well before the discovery of DNA and its mechanism. Though law has certain exceptions, still it serves as a rule of thumb to determine relative radiosensitivity of the cells. Mitotic rate is one of the important factors, which influence the radiosensitivity of cells. It is found that rapidly dividing cells are killed with relatively low radiation exposure. Based on the criteria of cell death, cells that have low mitotic rate are more resistant. Certain types of damage are produced during the irradiation, but cells die only when the cells enter division. If the mitotic rate is high, there will not be enough time for the cellular lesions to be repaired.

The cell with long mitotic future is the one which can undergo many cell divisions. In normal circumstances, the liver cells do not multiply, but if a small part of the liver is cut off, the cells multiply and the liver is regenerated. On the other hand, if the nerve cells are damaged they cannot be regenerated. Hence, liver cells have a greater mitotic future than the nerve cells and are more sensitive towards radiation. Mammalian cells, which are less differentiated, are more radioresistant than the highly specialized cells. However, there are exceptions to the above law, as seen in lymphocytes, neoplasms and oocytes. Though they are non-dividing cells, they are most radiosensitive. Based on radiosensitivity, tissues are divided as given below with decreasing order:
- Vegetative intermitotic cells, e.g. erythroblasts
- Differentiating intermitotic cells, e.g. myelocytes
- Reverting postmitotic cells, e.g. liver
- Fixed postmitotic cells, e.g. nerve cells.

Effects on Cell Cycle

Cells are more sensitive during M and G2 phase of the cell cycle, and more resistant during the S phase. Experiments have shown that there is a depression of DNA synthesis, after radiation exposure. This is mainly due to G2 arrest, which reduces the number of cells moving around the cycle and entering the S phase. The extent of radiation-induced depression of DNA synthesis is proportional to the oxygen tension. It is also found that cells are most sensitive for 3 hours proceeding mitosis, which corresponds to G2 period. In general, radiation slows down all stages of the cell cycle; the most important delay is the G2 arrest, which results in division delay. The magnitude of division delay is dose dependent. Further, cells irradiated in G2 phase show a greater delay compared to cells irradiated in S phase, suggesting that most of the division delay is caused by the inhibition of mitosis. During the delay period, the cells may be repairing the damage induced by radiation, which may be a very important requirement of the cells to survive the crucial test of mitosis.

The rapidly dividing cells spend more time in M phase; if the MI is higher, there is more possibility of cell death. The sulfhydryl compounds are the natural radioprotectors.

Their intracellular level is related to radiosensitivity, since the level vary with cell cycle phase. In the S phase, the DNA is replicated and all biochemical compounds for replication and repair are available. Hence, it is suitable site for repair and all the radiation damages are repaired. Therefore, it is most radioresistant. On the other hand, the M phase involves series of steps that includes condensation, alignment, attachment of spindles and separation of sister chromatids. Failure of any one step, make the cell fail to divide. Thus, the radiation damage has greater potential to kill the cell in M phase, compared to other phases.

Research reveals that the cell cycle progression is controlled by a family of genes called molecular checkpoint genes. The function of checkpoint genes is to ensure the correct order of cell cycle events. A specific gene halts the cells in G2, so that the inventory of chromosome damage is taken, repair is initiated and completed before mitosis. A mutant cell, does not have the above G2 checkpoint gene, moves directly into mitosis and has higher risk of dying.

Alternatively, these types of cells are more sensitive to radiation, since they have damaged chromosomes. The checkpoint controls and monitors the spindle function during mitosis. If the spindle is disturbed, the progression through mitosis is blocked. The G2 checkpoint gene involves cyclin-dependent kinase 1 (Cdk1), i.e. p32 protein kinase. The levels of the gene control the passage through mitosis. The cells that lack checkpoint genes are also prone to carcinogenesis. This G2 block causes inverse dose rate effect at low-dose rates. That is, as the dose rate decreases, the cells are more radiosensitive as seen in the range 0.3–0.4 Gy/h in brachytherapy.

Giant Cell Formation

Cells survive after a high dose of radiation above 10 Gy, generally form colonies with most of the cells considerably larger than normal cells. Cells exposed to < 10 Gy, generally form colonies containing both normal cells and large cells called the giant cells. Giant cells vary in size and sometimes grow 100 times larger than normal cells. Protein and nucleic acid content of giant cells are proportional to their volume. Hence, formation of giant cells is due to some difficulties associated with cell cycle division. Inhibition of cell division and endoreduplication (synthesis in the absence of cell division), and several sequences of progression in the absence of mitosis are suggested as the cause for giant cell formation. Finally, the giant cell undergoes death by lysis.

SYNCHRONOUS CELL CULTURES

In general, cell population is asynchronous during irradiation, i.e. the cells are distributed throughout in all the phases of the cell cycle. To study the radiosensitivity of the cell cycle, the cells must be made synchronous; they should be in the same phase of the cycle, at a given time. There are two methods by which cells can be made as synchronous population namely:
- Mitotic harvest
- Use of drug.

The mitotic harvest is used only for cultures that grow monolayers attached to the surface of the growth vessel. When the cells are close to mitosis, they round up and get loosely attached to the surface. If the growth medium is shaken, the mitotic cells detach from the surface and float in the medium. Now, the medium is removed from the culture vessel and plated in a new petri dish. The dish now consists of only the mitotic cells that are incubated at 37°C. This causes the cells to be synchronous and moves together through the mitotic cycle. The hydroxyurea is the most commonly used drug for making synchronous population of cells, which can be used both in culture and in tissue. If the

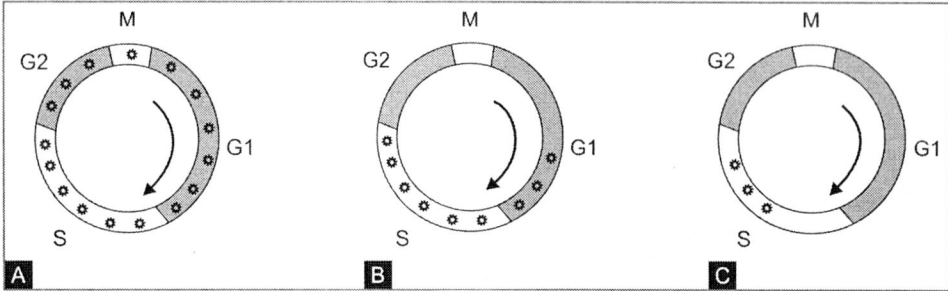

FIGURES 3.6A to C: Formation of synchronous cell population by the drug hydroxyurea. **A.** The drug kill the cells in S phase and imposes block at the end of G1 period; **B.** Cells at the G1, M and G2 phase gets accumulated at the block; **C.** Block is removed, the synchronous cells move through the cell cycle (G1/G2, interphase; M, mitosis; S, synthesis).

drug is added to the population of dividing cells, it has two effects (Figs 3.6A to C):
- The cells in the S phase are killed
- The drug impose block at the end of G1 period.

The cells in the G2, M and G1 phase progress through the cell cycle and get accumulated at this block. The drug is left in the cell for a period equal to the combined period of G2, M and G1. At the end of the period, all the cells, which are ready to enter S phase, are brought to the end of G1. If the drug is removed, then the synchronous cells progress through the cell cycle.

RADIATION AND CHROMOSOME DAMAGE (Figs 3.7A to E)

Irradiation can damage chromosomes and delay the cell's entry into mitosis. This delay is dose dependent. If the cells are irradiated during the interphase, they begin to divide and undergo aberrations. Chromosome aberration refers to their appearance in the first metaphase after irradiation. The stable aberrations may be carried out through number of cell divisions, generally called chromosome aberrations. Unstable aberrations may lead to cell death. If cells are exposed to radiation, DSB occurs in chromosomes. The broken ends are sticky due to their unpaired bases. These broken ends may:
- Join with their original chromosome
- Fail to rejoin, which leads to aberrations
- Rejoin with other broken ends.

These aberrations may be chromatid or chromosome aberrations. Chromatid aberrations are due to irradiation of cells, while they are in late interphase (G2). In this, one arm of the sister chromatid is broken. The DNA has already doubled and the chromosomes consist of two strands of chromatin, during irradiation. If the cells are irradiated, while they are in early interphase (G1) and if unrepaired, leads to chromosome aberrations. In this case, the chromosome is not duplicated and the damage is a SSB of chromatin. During the S phase, an identical strand is synthesized. This is visible in the mitosis as identical breaks in the pair of chromatin strands. Irradiation of cells, while they are in S phase may bring both types of aberrations.

Lethal Type Chromosome Aberrations

The types of chromosome aberrations that are lethal to cells are:
- Dicentric
- Ring
- Anaphase bridge.

Lethal Type Chromatid Aberrations

Anaphase Bridge

Anaphase bridge formation is the result of radiation-induced breaks at the later part of the interphase (G2), after replication. In this case, the break occurs in each chromatid of the same chromosome. The damaged sister chromatids do not separate properly, but stretch to the extreme poles of the cell. The broken ends join incorrectly and form a sister union. This results in a dicentric chromatid known as anaphase bridge. This type of aberration prevents the separation of parent into daughter cells, resulting in reproductive death. In addition, a fragment is formed without centromere, which will be lost during mitosis. Anaphase bridge type of aberration is usually lethal to the cell.

Non-lethal Type Aberrations

The types of aberrations that are not lethal to cells are:
- Symmetric translocation
- Small interstitial deletion.

They basically arises from DSB, remains intact with two chromatids and centromere, and carryout mitosis normally. However, there may be some loss of genetic information that may be passed on to the next generations.

Symmetric Translocation

In symmetric translocation, breaks occur in two different chromosomes at the early interphase (G1). The broken ends are exchanged between two chromosomes and get rejoined. This type of aberration can be easily seen with fluorescent in situ hybridization (FISH) technique or chromosome painting. Using the FISH technique, translocations can be scored on the exposed individual, even after 50 years of radiation exposure. Thus, translocations

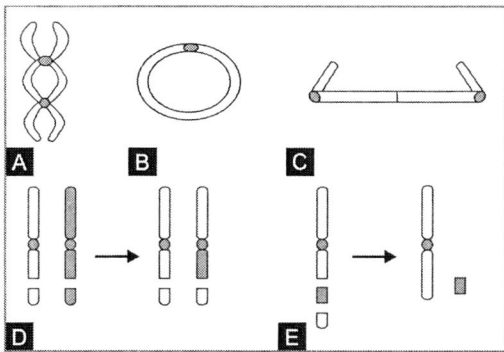

FIGURES 3.7A to E: Radiation-induced chromosome damage. A. Dicentric; B. Ring formation; C. Anaphase bridge; D. Symmetric translocation; E. Small interstitial deletion.

Dicentric

Two separate chromosomes undergo breaks at the early interphase. If the broken ends are close to each other, they get united illegitimately and replicated during the S phase. After the synthesis, a pair of sister chromatids appears with two centromeres, which is called dicentric. It is an unstable aberration lethal to the cell, not passed on to the progeny. It decreases slowly over a period of time after radiation exposure. Since spindle has two centromeres to grab during the metaphase, it gets disturbed, while pulling the chromatids during the anaphase.

Ring

Radiation induces breaks in each arm of a single chromatid, at the early part of the cell cycle. The breaks may rejoin and form ring and fragments. During synthesis, it gets replicated and appears as overlapping rings and acentric fragments without centromere. Since the ring or the acentric fragment do not have properly constructed centromere for the spindle to attach, chromosomes are unable to divide properly. The fragments will be lost during next mitosis.

are stable aberrations, present for several years and are passed on to the progeny. It can lead to mutation and carcinogenesis, but they are not lethal to the cell.

Small Interstitial Deletion

In small interstitial deletion, two breaks occur in the same arm of a single chromosome. The sticky ends may rejoin with a deletion with loss of DNA. The deleted part may be a form of acentric ring without centromere, which will be lost during next mitosis. This type of aberration may lead to malignancy in humans.

In general, cells can tolerate a variety of structural changes. The individual can exist with changes throughout the life. Sometimes, chromosomal instability in structure and magnitude may produce tumors. Irradiation can enhance the above process. Basically, cell death arises from chromosome aberrations, where there is a partial loss of genome. In a diploid cell, formation of micronucleus gives rise to cell death. In the case of polyploid cells, it is not that much serious, since multiple copies of chromosomes are present.

BIODOSIMETRY

Biodosimetry is one of the methods of measuring radiation-absorbed dose in human. There are several physical and chemical methods that are available for the estimation of radiation-absorbed dose in human, especially in radiation workers. The other methods are not very much useful under accidental radiation exposure due to their nonavailability and also it may be partial body exposure. Hence, under accidental condition, it is the biological dosimeter that plays an important role in the estimation of radiation-absorbed dose, as this method uses the biological sample such as blood, obtained from the exposed individual.

During radiation accident, various organs of the body receive radiation damage and they show different effects based on the severity of the damage. At cellular level, it should be understood that any cell will suffer chromosomal damage. For biological dosimetry purpose, the cell to be assayed should be matured with a long residence time in the body and should be easily obtained from the body. The small lymphocyte present in the blood is highly suitable for this purpose. The cell is in G0 stage of the cell cycle and undergoes no further cell division when mature and distributed throughout the body. Hence, it is ideal to collect these cells for dicentric assay.

Chromosomal Aberration Analysis

Chromosome aberration occurs spontaneously. Radiation-induced chromosome aberrations in human lymphocytes can be used to quantify radiation exposure in radiation accidents. The extent of genetic damage depends on the:
- Cell type
- Number and kind of genes deleted
- Either somatic or germ cell.

Chromosomal aberration analysis is based on the measurement of the frequency of dicentric chromosome. Blood sample is collected within few weeks of whole-body irradiation from the exposed individual and culture the lymphocytes in growth medium containing antibiotics and phytohemagglutinin (PHA) for 48 hours at 37°C. The PHA is added to stimulate the cells to divide. At 45 hours, colcemid is added to arrest the cells at metaphase. At 48 hours, the culture is centrifuged and the cells are suspended in 0.075 M potassium chloride (KCl) for 10 minutes to provide hypotonic treatment. The cells are then resuspended in fixative methanol:acetic acid (3:1). The cells are washed again with the above fixative and the slides are prepared and stained with fluorescent dye and Giemsa. The cells are then scored for chromosomal aberration analysis. Cells having aberrations such as dicentrics and rings are carefully noted. For biological

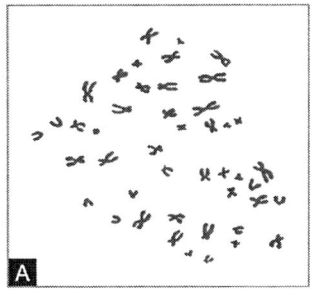 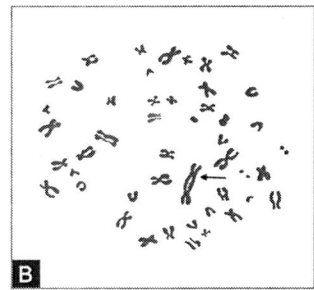 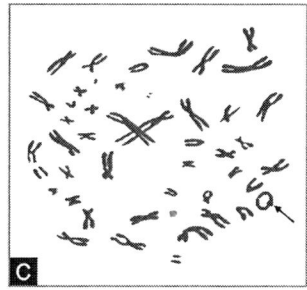

FIGURES 3.8A to C: Spread of chromosome seen under microscope. **A.** Normal metaphase chromosome obtained from the human peripheral blood lymphocytes; **B.** Metaphase chromosome with a dicentric (arrow) obtained from the human peripheral blood lymphocytes exposed to gamma (γ) radiation; **C.** Metaphase chromosome with a ring (arrow) obtained from the human peripheral blood lymphocytes exposed to γ-radiation.

dosimetry purpose, several hundred cells are scored to get the right estimation of dose. The scoring is studied with a standard dose-response curve to estimate the radiation dose.

It is important that the laboratory carrying out the biological dosimetry work has to construct its own dose-response curve. The dose-response curve is prepared by exposing in vitro blood samples with different radiation doses. The aberration per cell for each dose is scored. Then, a graph is fitted between absorbed dose (Gy) and aberrations per cell, in X- and Y-axis respectively. The frequency of chromosome aberration is a linear quadratic function of dose for low-LET radiation. With regard to biological indicators, although various methods are available, the gold standard for estimation of radiation dose is the chromosomal aberration technique by way of measuring the frequency of dicentric chromosome. The Figures 3.8A to C shows the spread of chromosome as seen under the microscope.

The lowest dose that can be detected by the above method is 0.1 Gy. Dicentric chromosome aberrations are unstable and may be lost from the blood over time. The life span of a lymphocyte is about 150 days, hence the dicentric yield decreases after a month or year of radiation exposure. This is a useful and most sensitive technique during overexposures or suspected exposures in personnel monitoring. The stable reciprocal translocations can be measured by FISH. In this, chromosome is labeled with chromosome-specific fluorescent DNA probes. Then, translocations can be viewed under the fluorescent microscope. Since reciprocal translocation exists for longer period of time, the FISH method is most suited for delayed investigations.

BIBLIOGRAPHY

1. Charles A Kelsey, Philip H Heintz, Danial J Sandoval, et al. Radiation Biology of Medical Imaging. Hoboken, New Jersey: Wiley Blackwell; 2014.
2. Eric J Hall, Amato J Giaccia. Radiobiology for the Radiologist, 7th edition. Philadelphia, USA: Wolters Kluwer/Lippincott Williams & Wilkins; 2012.
3. Jerrold T Bushberg, Anthony Seibert J, Edwin M, et al. The Essential Physics of Medical Imaging, 3rd edition. Philadelphia, USA: Wolters Kluwer/Lippincott Williams & Wilkins; 2012.
4. Johns HE, John R Cunningham. The Physics of Radiology, 4th edition. USA: Charles C Thomas; 1984.
5. Rao BS. Interaction of Radiation on Cells. In: One Year Post Graduate Course in Hospital Physics and Radiological Physics. Mumbai: Bhabha Atomic Research Centre; 1984.

Cell Survival Curves

RADIATION AND CELL DEATH

Radiation exposure to mammalian cell population results in two types of cells namely:
- Dead cells
- Surviving cells.

Dead cells are the one, which have lost the capacity to divide indefinitely, i.e. loss of reproductive integrity. The cells, which are physically present, synthesize deoxyribonucleic acid (DNA) or make proteins, but not able to divide during mitosis, are also said to be dead. Cells that have retained its reproductive capacity and are able to divide indefinitely are called survived cells. These cells can produce colony, and hence, said to be colony-forming cells. Death in differentiated cells leads to loss of organ function, e.g. nerve, muscle, etc. The cell death in a proliferating cell (stem cell) leads to loss of reproductive capacity, e.g. blood cells, epithelium, etc. Cells generally die in two ways:
1. They die when they attempt to divide; it is called mitotic death.
2. Cells which undergo a programmed death; it is called apoptosis.

The relation between the radiation dose and the number of surviving cells that form colonies can be plotted in a curve known as survival curve. The survival curve is a measure of reproductive death. To conduct the experiment, a known number of cells are seeded in a petri dish (in vitro) and exposed to varying doses of radiation. Usually, multiple petri dishes with cells are seeded. One or two dishes are used as control and no radiation is given. The irradiated and the control cells are incubated for same duration of time. They are allowed to grow colonies and the same is counted. Each cell is assumed to make its own colony. The fraction of surviving cells is normalized by the fraction of cells that survive with no radiation exposure.

A graph is plotted between the radiation dose and the survival fraction. The survival curve of the mammalian cells is exponential in nature, and hence, plot is usually drawn between log of survival fraction (Y-axis) and radiation dose (X-axis). These curves are useful to understand the sensitivity of cells to radiation and their nature. The shape of the curve tells us the radiosensitivity, repair and recovery ability of the cell. The shape of the cell survival curve gets altered, if the exposure conditions are changed. To quantify the cells, the parameters used are:
- Plating efficiency (PE)
- Survival fraction (SF).

SURVIVAL FRACTION AND PLATING EFFICIENCY

The surviving fraction is given by the relation:

$$SF = \frac{\text{Colonies counted}}{\text{Cells seeded} \times \dfrac{PE}{100}} \quad \text{-----(1)}$$

where,

PE is the plating efficiency, which refers to the percentage of seeded cells that survive to form colonies under controlled conditions:

$$PE = \frac{\text{No. of colonies counted}}{\text{No. of cells seeded}} \times 100 \quad \text{-----(2)}$$

The PE can be obtained with in vitro studies as follows. The cells are prepared in a culture vessel (dish) with trypsin and the number of cells are counted by an electronic or hemocytometer. The dish is incubated for 1–2 weeks. During this period, the cells divide and forms colonies. Now, the number of colonies is again counted using the above formula to get the PE. The number of colonies is always lesser than the number of seeded cells due to growth medium, uncertainties of counting and trauma of trypsinization and handling.

SURVIVAL CURVE

In vitro studies show that survival curve of each cell line after radiation exposure can be plotted in a graph with the surviving fraction in the Y-axis and the dose in the X-axis (Figs 4.1A and B).

The curve in Figure 4.1A plotted on a linear graph contains two regions namely, region A at low doses and region B at low survivals, both are important for radiotherapy. However, the curve may not be useful clinically as it is not giving the true picture of cell survival. In radiotherapy, we use low dose per fraction (1–3 Gy per fraction) and tumor cure require cell survival of about 10^{-8} cells. Hence, it is meaningful to plot the curve as shown in Figure 4.1B in a semilog scale. This curve demonstrates the exponential behavior of region A as straight line. It also magnifies the region B, which represents the low survival levels of cell (10^{-8}) and helps to visualize tumor cure.

Survival Curve Characteristics

The cell survival curve is exponential in nature, if it is plotted in linear graph. If it is plotted in a semilog graph with log of SF in the Y-axis and radiation exposure in X-axis, it appears as straight line with an initial shoulder in case of mammalian cells. This suggests that the efficiency of cell killing is lower at low doses, which is due to the ability of cells to accumulate or repair such sublethal damage (SLD). The shape of survival curve varies with linear energy transfer (LET) of the radiation. Generally, low-LET radiation (X- and γ-rays) curve is taken for discussion, since it is commonly used in medicine. To explain the survival curve, several theoretical models have been proposed namely:

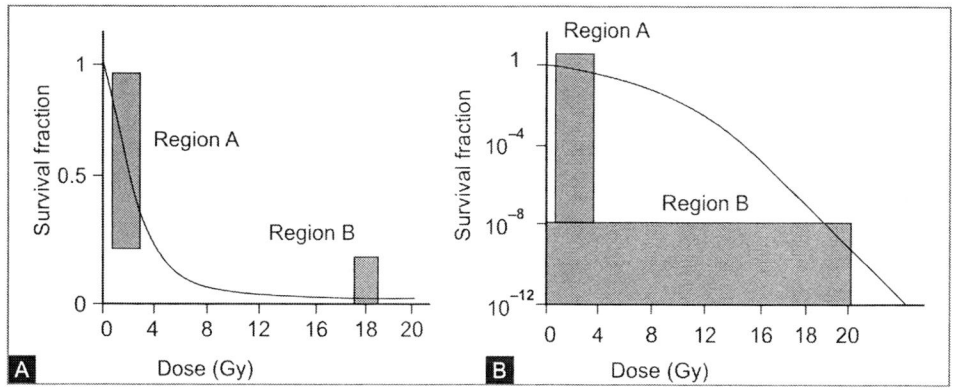

FIGURES 4.1A and B: Shape of the survival curve for mammalian cells. **A.** Survival fraction and dose plotted in linear scale; **B.** Survival fraction plotted in logarithmic scale with dose on a linear scale.

- Single target-single hit theory
- Multitarget theory
- Linear quadratic (LQ) theory.

The target theory gives the relationship between the number of cells killed and the radiation exposure they received. This was proposed by Growther and then expanded by Lea. The target is the critical site in the cell, which must be damaged in order to kill the cell. The production of ionization in the target volume is termed as hit. The shape of the survival curve is governed by:

1. Number of targets within each cell that have to be damaged.
2. Number of times each target needs to be hit before the cell loses its ability to reproduce.

However, no theory completely explained the survival curve that is obtained experimentally for the mammalian cells.

Single Target-single Hit Theory

The single target theory assumes that each cell has single target and if the target receives a single hit, the cell will die. The lethal event is one where a single ionizing particle interacts with a single target. The cell has no opportunity to repair the radiation damage. Such events are random in nature, since there is very large number of cells in an organism. Hence, the surviving fraction (S) is described by Poisson statistics, as follows:

$$S = \exp(-P) \quad \text{-----（3）}$$

where,

P is the mean number of probability for not having a hit per cell. If the probability is a linear function of dose (D), then the equation can be written as:

$$S = \exp\left(-\frac{D}{D_0}\right), \text{ taking logarithm}$$

on both sides

$$\ln S = -\frac{D}{D_0} \quad \text{-----（4）}$$

where,

D_0 is the mean lethal dose that will give one hit per target. If $D = D_0$, then $S = \exp(-1) = 1/e = 0.37$.

D_0 is the reciprocal of the slope of the linear portion of the survival curve. This dose will bring the survival fraction to 37% (Fig. 4.2). Hence, D_0 is also referred as D_{37}. D_0 is higher for radioresistant cells and lower for radiosensitive cells. The typical value for mammalian cell is 1–2 Gy for low-LET radiation. At low doses, the probability of producing more than one hit is small, and hence, it follows a linear relation. However, at larger doses, the probability of ionization occurring in already inactivated cells will be higher and hence, it is wasted, since one hit only is needed to kill the cell.

However, the single target theory fails to explain the experimental survival curves, which are obtained for various cell population. It assumes that:

1. The dose rate is independent of observed biological effect.
2. Experimental conditions during and after irradiation has no influence.
3. High-LET radiations are less effective, which are not true.

In the case of mammalian cells, the linear characteristic of survival curve is obtained for high-LET [alpha (α) particle] irradiation. This model does not account the shoulder region

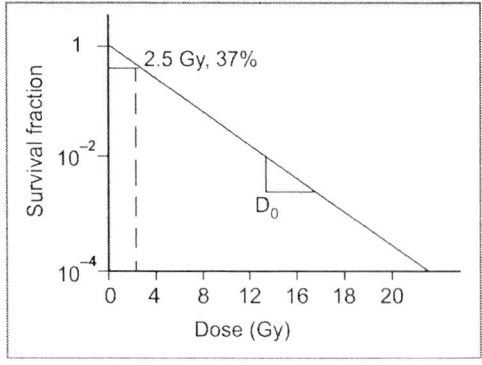

FIGURE 4.2: Survival curve of single target theory in a semi-log scale. It is a hypothetical curve drawn with a mean lethal dose of 2.5 Gy.

of the curve at low doses, and hence, has little application in practice. This may be useful to describe viruses, but not mammalian cells.

Multitarget Theory

Single hit-multitarget theory

Multitarget model is also known as two-component model. According to this theory, some cells contain more than one target and each of the targets should receive the hit, i.e. this theory assumes the concept of single hit, multitarget model. The survival curve starts with less sensitive region at low doses and then tends to become exponential at large doses (Fig. 4.3). Due to this reason, they present a picture of shoulder at smaller doses, hence termed as sigmoid or shoulder type survival curves.

There are 'n' distinct targets in each cell and all the targets should receive at least one hit within each cell. The probability for no hit in the target is:

$$\exp\left(-\frac{D}{D_0}\right)$$

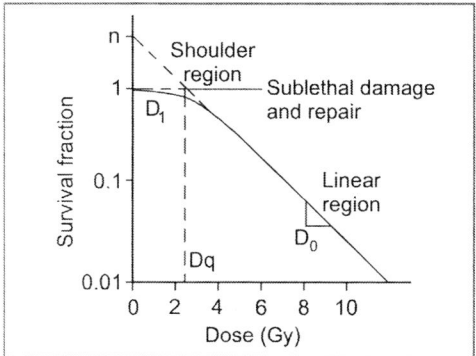

FIGURE 4.3: Survival curve for mammalian cells exposed to X-rays in a single hit-multitarget model. D_1 and D_0 are the initial and final slope. D_q is the quasi-threshold dose, which refers to the width of the shoulder and n is the number of targets.

The probability that there is at least one hit per target is:

$$1 - \exp\left(-\frac{D}{D_0}\right)$$

If there are 'n' targets in the organism, then the probability of hitting once all the targets is:

$$\left[1 - \exp\left(-\frac{D}{nD_0}\right)\right]^n$$

Then, the SF of cells is given by equation:

$$S = 1 - \left[1 - \exp\left(-\frac{D}{nD_0}\right)\right]^n \quad \text{-----(5)}$$

A survival curve is defined with set of parameters, namely:
- Initial slope (D_1)
- Final slope (D_0)
- Extrapolation number (n)
- Quasi-threshold dose (D_q), as per the multitarget theory.

Here, D_1 is the reciprocal of the initial slope called mean lethal dose. It is the dose required to reduce the survival to 37% of its initial value. That is the surviving fraction reduces from 1 to 0.37, or 0.1 to 0.037, and so on. Similarly, D_0 is the reciprocal of the final slope and the dose required to reduce the survival to 37%. In general, D_1 and D_0 are the average doses required to create at least one inactivating event per cell.

The shoulder represents those cells in which less than the required number of targets have been damaged for a given dose. These cells are said to have received SLD. These cells are capable of repairing their damage, which is called SLD repair. Repair is a time-dependent phenomenon and typical repair half-time is of the order of 1 hour. Complete repair of SLD occurs between two dose fractions delivered with an interval of about 6 hours apart.

If the straight line portion is extrapolated, it meets the Y-axis at a point corresponding to the number of targets 'n' called extrapolation number. It is a measure of the width of the shoulder. If 'n' is large, the width of the shoulder is broad and smaller 'n' refers narrow shoulder. The extrapolation number of mammalian cell varies between 2 and 10. If n = 1, the equation 5 is reduced to equation 4, the multitarget model reduced to single hit-single target model. However, if different LET radiations are used for a given cell line, one can get different 'n' values, i.e. 'n' value changes with the type of radiation, which is much controversial.

The D_q is the width of the horizontal line parallel to X-axis drawn at the survival level of 1. It reflects the ability of the cells to accumulate SLD, a measure of shoulder width. SLD is a parameter obtained from experiments. If the radiation dose is split into fractions with sufficient gap between fractions, the shoulder reappears. This confirms the theory of SLD that:

1. More than one ionization event is required to kill the cell.
2. The cell is capable of repairing damage.

However, this model can be used for both low- and high-LET radiations.

Single hit-multitarget and single hit-single target theory

The earlier single hit-multitarget theory explained nicely the shoulder region of the mammalian cells; however, the cause of initial slope is not explained. Hence, it is believed that there are two independent events that are involved in cell inactivation. One is the single hit-multitarget mechanism that requires at least one hit for each target. This is responsible for the accumulation of SLD in the shoulder region, which is explained earlier. The second event is caused by the passage of single charged particle. It hits all the targets simultaneously and is believed to be single hit-single target concept. Hence, the single hit-multitarget and single hit-single target, both events are responsible for cell killing in mammalian cells (Fig. 4.4). The former event is a repairable damage, whereas the latter damage is an irreparable damage. Thus, radiation causes both repairable and irreparable events. The survival fraction for these types of events is obtained by multiplying single hit-single target and single hit-multitarget equations:

$$S = \exp\left(-\frac{D}{1D_0}\right)^1 \times \left\{1 - \left[1 - \exp\left(1 - \frac{D}{nD_0}\right)\right]^n\right\}$$

------ (6)

where,

$1D_0$ is the mean lethal dose for single direct lethal events; nD_0 is the lethal dose for sublethal events.

Systematic analysis of the curve reveals an initial straight line with finite slope, a bend with shoulder and again a straight line with a deeper slope for a low-LET radiation, i.e. it starts and ends with an exponential relation between SF and radiation dose. If the $D > D_q$, the relation between the dose and survival fraction is a straight line with a slope of $1/D_0$. If D_0 is small, the slope is high and the cell is said

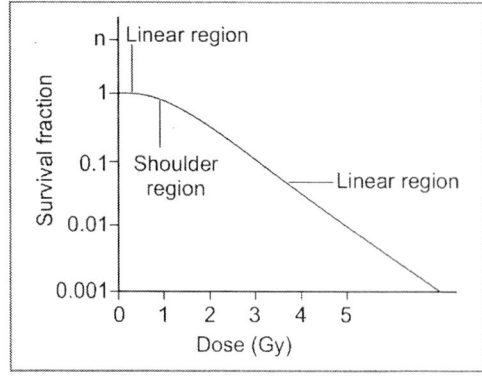

FIGURE 4.4: Single hit-multitarget and single hit-single target survival curve

to be more radiosensitive. On the other hand, if D_0 is large, slope is less and the cell is more radioresistant. This may need large additional dose to increase the cell killing.

Survival curve parameters are useful to compare radiation response under different physical conditions. The survival curve shape varies between low- and high-LET radiations and between cell lines. High-LET radiation survival curve exhibits absence of shoulder, hence, n and D_q have no relevance. Mammalian cells have values of D_0 = 1–2 Gy, n = 1–5 and D_q = 0.5–2.5 Gy, for low-LET radiation. However, when cells exposed to low-LET radiation in the presence of certain agents can modify survival curves.

Linear Quadratic Theory

Linear quadratic (LQ) theory states that inactivation of a cell results only when both strands of DNA molecules or both arms of a chromosome are damaged. This can be produced either by passage of single ionizing particle across the cell or by independent interactions of two separate ionizing particles.

Thus, the LQ theory assumes that there are two ways of cell killing by radiation, namely linear and quadratic (Fig. 4.5). This can be explained in terms of physical interactions of radiation on cells. Double-strand break (DSB) in DNA molecules are the lesions mainly responsible for cell lethality. The linear part of DSB is caused by the passage of single ionizing particle. The probability of such event can be represented by Poisson statistics. Hence, the mean number of such events is 'αd', where α is the probability per unit dose that such an event will occur.

The 'α' refers to irreparable damage and is equal to $1/1D_0$ in single hit-single target theory. The probability of no such events (surviving fraction) is given by:

$$S = \exp(-\alpha d) \quad \text{-----(7)}$$

The quadratic portion of DSB is caused by independent interactions of two separate ionizing particles. For each particle, the probability is linearly related to dose. The mean probability of such events for two particles is βd^2, where 'β' is the average probability per square of the dose. The probability that no such two particle events will occur is given by:

$$S = \exp(-\beta d^2) \quad \text{-----(8)}$$

This represents the repairable damage. Combining both the ways of DSB cell killing the total events are:

$$\alpha d + \beta d^2$$

The total probability that no such events or SF of cell after radiation exposure, is given by the Poisson statistics as:

$$S = \exp - (\alpha d + \beta d^2) \quad \text{-----(9)}$$

By taking natural logarithm on both sides, the equation can be written as follows:

$$-\ln S = \alpha d + \beta d^2 \quad \text{-----(10)}$$

The LQ theory describes the cell killing in a better way in the low-dose range (0–3 Gy). As per the theory, the curve is continuously bending without initial and final straight line portion. The shape of the curve is determined by the α/β ratio, which is a dose and its unit is Gy. The term 'α' is the coefficient of cell killing that is proportional to dose, whereas 'β' is the coefficient of cell killing that is proportional to the square of the dose.

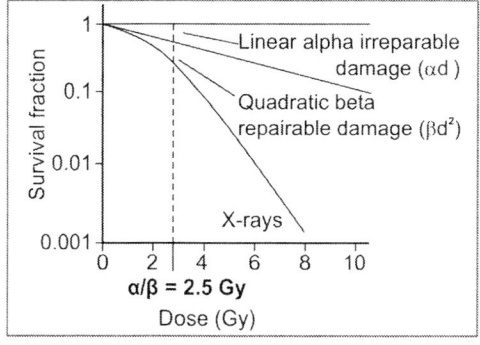

FIGURE 4.5: Survival curve and the linear quadratic model for mammalian cells exposed to X-rays. The dose at which the linear and quadratic components are equal is the ratio of α/β.

The DSB may be caused by either single radiation track or double radiation tracks. The breaks must be close to each other in terms of time and space. The α-coefficient represents the single radiation track, which is dose rate independent. The damage is irreparable over time and mostly dominates in high-LET radiations. The β-coefficient represents multiple radiation tracks; its damage is repairable and depends on dose rate. It is responsible for the bending of survival curve at higher dose and dominates in low-LET radiations. The ratio of α/β is a dose at which the linear part of cell killing is equal to quadratic part of cell killing, i.e.

$\alpha d = \beta d^2$ or $d = \alpha/\beta$.

The α/β represents the curviness of the cell survival curve. It is another way of characterizing the shoulder. Higher the α/β, or higher the α, or lower the β, straighter the survival curve and narrow the shoulder. It is a characteristic of a type of cell that exhibits irreparable damage and capable of little repair of SLD. It indicates that the tissue is radiosensitive or early reacting tissue. At low α/β ratio, or higher β value, or lower value α, it refers wide shoulder, high capability of repair and indicates little irreparable damage. It refers the tissue is radioresistant or late responding tissue. There exists a major difference between cells of tumors and those of late responding normal tissues. Tumor tends to have high α/β values, whereas it is much lower for normal tissues. For example, typical values determined for cancer cells range from 8 to 12 Gy (≈ 10 Gy). The corresponding range for late responding normal tissue is 2–4 Gy (≈ 3.0 Gy) and early responding normal tissue is 6–12 Gy.

Relevance of Survival Curves

Single hit-single target theory supports the mammalian cells under certain conditions. Especially with high-LET radiations or synchronized cell populations in the sensitive phase of the cell cycle. Multitarget survival curve fit very well at high-dose region and can be used for both low and high-LET radiations. It can be used to study radiosensitivity of the cell with different types of radiations. However, it is not supporting for low-dose region, especially the shoulder regions.

The LQ model is suitable and accurate for low-dose region and clinically much useful. With this model one can study and compare the radiosensitivity of different cell lines. The LQ model is useful in radiotherapy to explain fractions and differentiate tumor, late and early responding normal tissues. Though it is not explaining the complete survival theory, it can be used to model low-dose survival curves. The LQ model does not support the high-dose region of the curve. However, the survival curves obtained under in vitro conditions are found to differ from that of in vivo conditions. Hence, extra care is needed, while using the survival curve information.

Factors Influencing the Survival Curve

The shoulder region is the most important factor in the survival curve, since it has relevance in clinical radiotherapy. It is helpful in predicting the effect of low doses in humans. In medicine, low-dose radiations are used for both treatment and imaging. The shape of the shoulder is different for different cell lines. On the other hand, the shape of the shoulder is different for a given cell line for different environment conditions. The shape of the shoulder is believed to have two assumptions:

1. Radiations, which are not individually fatal, may be collectively fatal.
2. Sites repairable, not fatal at low doses and fatal at higher doses.

In the former case, a high-LET radiation; however, can offer multiple hits to kill the cell in a single event resulting in no shoulder. In the case of low-LET radiation delivered in fractions, it has shoulder, since the damage of the first fraction is accumulated.

In the latter case, repair of damage is possible at lower doses, but it is not possible at higher doses as the cell is killed due to more number of ionizing events take place in the cell. In multiple hit, there may be delay between hits. If the cell repair the damage and present during the second fraction, the shoulder will appear broader. However, there are limitations on the rate of DNA repair and there are possibility that repair is not completed before the second fraction. If the second fraction dose is high, it exploits the situation and exhibit narrow shoulder.

REPAIR OF RADIATION DAMAGE

Cells generally try to restore the integrity of the DNA molecule before they enter into the first postirradiation cell division. Availability of repair mechanisms decide the extent of damage that could be repaired, and hence, the radioresistance. It is possible to repair the radiation damage including base damage, SSB, DSB, sugar damage and DNA-DNA crosslink. The pathways used to repair the base damage are different from that of DSB damage. In addition, the repair pathway varies with the stage of the cell cycle. The repair of damage depends on stage of the cell cycle, the type and location of the damage.

There are specific endonucleases, exonucleases and proteins that repair the DNA damage. In general, the cell activates G1/S checkpoint, arrests the cell in the cell cycle and completes the repair. In potential mutations, the cell may initiate apoptosis and eliminate the damaged genetic material. Apoptosis is an event of programmed cell death, which is different from necrosis. This leads to cell shrinkage by nuclear condensation and membrane bleeding. This will result in fragmentation of the cell into membrane-bound apoptotic bodies composed of cytoplasm and tightly packed organelles, which are eliminated by phagocytosis. The various DNA repair pathways are:
- Base excision repair
- Nucleotide excision repair
- DNA double-strand break repair
- Crosslink repair
- Mismatch repair.

Base Excision Repair

The base excision repair (BER) is based on the mechanism that bases on the opposite DNA strands are complementary to each other. Most of the SSB is repaired by this pathway within hours of damage, i.e. adenine (A) pairs with thymine (T) and guanine (G) pairs with cytosine (C), etc. Base excision repair is possible both in single and multiple nucleotides. In the single nucleotide repair, the mutated single base (U) is removed along with sugar residue and it is replaced with the correct nucleotide. In the case of multiple nucleotides, the mutated bases (UU) are removed and the repair synthesis is performed and replaces the missing nucleotide. Finally, overhanging flap structure is removed and the DNA strand nick is sealed by ligase.

Nucleotide Excision Repair

The nucleotide excision repair (NER) is carried out in two ways, namely global genome repair (GGR) and transcription-coupled repair (TCR). It is a major pathway of repair of bulky, helix-distorting lesions, e.g. thymine dimmers by ultraviolet (UV) radiation. In GGR, both the lesions with encoded DNA genes and unencoded DNA genes are removed. The TCR pathway removes only lesions with DNA-transcribed genes. It removes the RNA polymerase, which are blocking the DNA-damaged site and facilitates protein to perform repair.

Thus, both the pathways of repair are the same, but their way of detecting the lesion is different. It starts with recognition of damage followed by DNA incision, which bracket the lesions. Then, the region containing the abduct is removed and the repair synthesis fill up the region. Finally, it ends up with DNA ligation.

DNA Double-strand Break Repair

Most of the DSB are repaired correctly, whereas few cells undergo misrepair. In the DSB repair, either homologous recombination (HR) repair or non-homologous end-joining (NHEJ) pathway is used. In HR, the homologous DNA (undamaged) is identified by the repair proteins. The undamaged homologous DNA and the damaged DNA have similar sequences. Therefore, it is a error-free process. The repair starts with nucleolytic resection of blunt ends and binding of Nijmegen breakage syndrome (NBS)/MRE11/RAD50 protein complex to the DNA termini. Then strand exchange takes place by attachment of RAD51/XRCC2 protein. This is followed by DNA synthesis of missing nucleotides using the undamaged templates and ligation. A complex strand crossover is created between the damaged and undamaged strands known as Holliday junction. This is finally resolved before the completion of repair process. Since DNA strand is used as template and repair is carried out with high fidelity, it is error-free, preserves genetic integrity, and occurs in late S and G2 phase of the cell cycle.

In NHEJ, DNA template is not used; instead, end-to-end joining of broken DNA fragments resulting from broken phosphodiester linkages is carried out. This starts with requirement Ku70/Ku80 repair proteins to recognize the lesion termini and binding of the Ku-heterodimer to DNA-dependent protein kinase (DNA-PK). This is followed by the activation of the XRCC4 ligase enzyme by this complex for final religation of the fragments after enzymatic cleaning up of the broken ends of the DNA molecule by a variety of other recruited proteins, so that ligation can occur. This type of repair takes place throughout the cell cycle, but dominates in G1/S phases. This is error-prone, since it does not rely on sequence homology.

Crosslink Repair

The crosslink damage of DNA-DNA and DNA-protein can be repaired. It uses both NER and recombinational repair pathways. The repair starts with removal of crosslink from one strand by NER and the resulting DSB is repaired by recombination.

Mismatch Repair

In mismatch repair, base-base mismatches are removed, which occurs during replication. It starts with identification of mismatch base pair, recruitment of mismatch repair factors, excision of incorrect nucleotides followed by a gap-filling synthesis and ligation.

Effects of Unrepaired Damage

The unrepaired DNA damage in chromatids can be detected by chromosome analysis (karyotyping), micronucleus formation and fluorescence in situ hybridization (FISH) techniques. The broken ends rejoin with other broken ends and form rings, dicentrics, translocations and other chromosome aberrations. The dicentric chromosome aberration arises from postreplication due to joining of two broken chromatids in different chromosomes and can be used as a marker for radiation damage. The acentric fragments and dicentrics are unstable aberrations and may not survive, if the cell tries to undergo cell division. This may lead to loss of genetic material and signaling death in diploid cells. Micronuclei contains acentric fragments and they can be detected by stimulating lymphocytes to undergo cell division and arresting the cells at cytokinesis using cytochalasin B treatment. It allows the nuclear division, but stops cellular division. Though micronuclei assay is less sensitive, it is a simple and alternative procedure for chromosome analysis.

MECHANISM OF CELL DEATH

The survival fraction decreases gradually with increasing radiation exposure. It indicates that there is heterogeneity in the cell population. Dose-response study reveals that, at low-dose range, the decrease in cell survival fraction is rapid with dose and at higher dose, decrease in survival is slow. Studies reveal that nucleus is the most sensitive site for cell death. DNA is the primary target for radiation-induced lethality. Cells are said to be killed, if they lost reproductive integrity, though they physically survive. The various mechanisms of cell killing are:
- Bystander effect
- Apoptosis
- Mitotic death
- Autophagic death.

Ultimately, this will lead to physical loss of cells, but take significant time to occur.

Necrosis occurs after high radiation dose. A rapid fall of cell number following irradiation is likely due to apoptosis, but may also occur due to mitotic catastrophe in rapidly proliferating populations. Whether apoptosis reflects overall cell killing in tumor cell inactivation by radiation, is uncertain and may only be the case for certain types of tumor cells.

Bystander Effect

When a cell population is irradiated, the cells surrounding the irradiated cell population are also affected and this effect is called radiation-induced bystander effect (RIBE). This effect is true for α, proton and soft X-rays. It is demonstrated using the studies such as chromosomal aberrations, cell killing, mutation, oncogenic transformation and alteration of gene expression, etc. It believes that the free radicals do not come in contact with cells and cause no radiation damage. Bystander effect is observed in both in vitro and in vivo experiments. It causes similar damages as caused by radiation such as DNA damage and reduced cell survival.

One of the important characteristics of bystander effect is genomic instability. Laboratory study with low-LET radiation reveals that the irradiated cells secrete molecule/chemicals into the medium, which are capable of killing cells. When the medium is transferred onto unirradiated cells, they establish bystander effect. Irradiation of Chinese hamster ovary cells with α-particles of dose less than 5 cGy and analysis of hypoxanthine-guanine phosphoribosyltransferase (HGPRD) mutations indicated very low track traversals of 0.05–0.3/cell, where, most cells are not hit, but the number of cells showing mutations was greater than the hit cells by a factor of 5. Similarly, irradiation of 1 human fibroblast on a dish with helium ion (He^+) from a microbeam produced 80–100 damaged cells and irradiation of the cytoplasm only produced DNA DSB in non-irradiated cells.

This suggests the role of gap junctions between cells to communicate damage-response signals. Alternatively, damaging molecules can be released into the medium surrounding the cells or the energy deposition in DNA is not required to trigger a bystander effect. The effect is more when cells are in the gap junction communication with irradiated cells. The effect is less when cell monolayers are sparsely seeded and cells are well separated. The effect indicates that the target is larger than the nucleus and it is very important at low doses. This has direct application in patients undergoing radiation therapy treatments.

Apoptotic Death

Apoptosis is a Greek word meaning 'falling off' that refers to programmed cell death. This involves sequence of morphological events. If the intended cell is to die, it initially rounds up and detaches from its neighbors, then their communication with neighbors ceases. This is followed by condensation of chromatin at the nuclear membrane and fragmentation of nucleus. Then, the cell shrinks due to cytoplasmic condensation as a result of protein crosslinking and loss of water. Finally, the cell is separated into several membrane-bound fragments of differing sizes known as apoptotic bodies. These apoptotic bodies contain either cytoplasm or nuclear fragments. Additionally, DSB occurs in the linker regions between nucleosomes resulting in DNA fragments and characteristic ladders.

Apoptosis is common in embryological development and lymphocyte turnover. Apoptosis occurs in normal cells and can be induced in tumor by radiation. In the case of tumor, apoptosis is cell type dependent. In most of the tumor cells, mitotic cell death is the radiation-induced mode of cell death. However, the hemopoietic and lymphoid cells are more prone to radiation-induced cell death through apoptotic mechanism. It can be identified by microscopy and typical shrinkage of cellular morphology, condensation of chromatin degradation, nucleosome laddering indicating chromatin degradation, cell membrane blebbing, activation of caspases and release of cytochrome c.

Radiation-exposed phosphatidylserine in the cell wall permits binding of annexin V and assessment of apoptosis by flow cytometry. The characteristics of apoptosis is in contrast to necrosis, typified by cell edema, poor staining of nuclei, increase of membrane permeability, shut down of cell metabolism and an accompanying inflammatory response. Replicative senescence is reported when cell stop dividing in contrast to the behavior of stem cells and tumor cells. Senescence cells are edematous and show poor cell-to-cell contact, increased polyploidy, decreased ability to express heat-shock proteins and shortening telomeres. Apoptosis occurs in particular in certain cell lines after low dose of radiation, e.g. lymphocytes, serous salivary glands, certain cells in testis and intestinal crypts.

Mitotic Cell Death

Reproductive cell death is a result of mitotic catastrophe, which can occur in the first few cell divisions after irradiation. It occurs with increasing frequency after increasing doses. Cells that fail to divide successfully after irradiation can also undergo apoptosis at that stage. The mitotic death is the most common radiation-induced cell death and it is due to chromosome aberrations. These aberrations are asymmetric exchange-type aberrations, e.g. ring and dicentric. Cells with asymmetric exchange-type aberration lose their reproductive integrity. Irradiated cells may undergo cell cycle and they may die in the first or second mitotic cycle, while attempting division.

The above type of aberration contains two chromosome breaks. At low doses, two such chromosome breaks are caused by single electron track. At higher doses, it is caused by two electron tracks as suggested by the linear quadratic theory.

Autophagic Death

Autophagy is a self-digestive process that uses lysosomal degradation of long-lived proteins and organelles to maintain cellular mechanism of cells. It is a protective mechanism for cells that generate nutrients and energy. However, stress-inducing condition can cause cell death known as autophagic cell death. Chemotherapeutic agents and radiotherapy can induce autophagy and autophagic cell death. Radiation combined with endoplasmic stress-inducing agents may enhance autophagic cell death.

CLASSIFICATION OF RADIATION DAMAGE

Radiation causes four types of damages in mammalian cells, namely:
- Lethal damage
- Potentially lethal damage
- Sub- and non-lethal damage.

The various sequence of radiation damage, repair and cell death is shown in Figure 4.6.

Lethal Damage

The lethal damage (LD) is an irreversible and irreparable damage, always leads to cell death. The cell may fail in its reproductive capacity, resulting in interphase and mitotic death.

Potentially Lethal Damage Repair

The potentially lethal damage (PLD) can be modified by postirradiation environmental conditions. The PLD damage is lethal under optimal conditions, but variation of postirradiation conditions can alter the damage and allow the cells to survive. Decreased survival is found under optimum conditions, since cells enter the mitosis without repair. PLD repair is possible if cells are prevented from division for 6 hours after radiation, which will increase the survival. This is proved in the in vitro experiments by:
1. Cooling the growth medium.
2. Adding dilute salt solution or density-inhibited state for 6–12 hours after radiation.

This creates a suboptimal condition for growth and prevents the mitosis and cell division. Thus, the PLD cells are allowed to rest and repair their radiation damage. If the repair is carried out correctly, the cell continues its function normally. If the repair is improper,

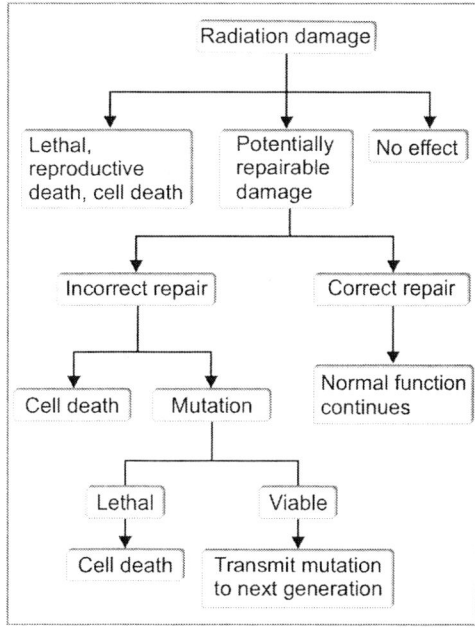

FIGURE 4.6: Sequence of possible radiation damage in cell

then the cell may either die or undergo mutation. The mutation leads to either cell death or viable mutation. PLD type of repair is significant for low-LET radiation. The existence of radioresistant tumors (e.g. melanoma) may be due to greater amounts of PLD repair. However, the role of PLD in clinical radiotherapy is yet to be justified.

Sublethal Damage Repair

The sublethal damage (SLD) can be repaired under normal conditions in hours after irradiation. The ability of the cell to recover from SLD has been demonstrated by Elkind and Sutton by split-dose experiments. A given total dose when delivered as a single fraction is more effective compared to the same dose delivered in two or more fractions separated by a time interval (Fig. 4.7). There is a steep increase in survival with increase in time duration and reaches a plateau at about 2 hours. The magnitude of surviving cells is four-fold compared to single dose. The rise in survival level during the first 2 hours is a result of the repair of SLD. As a result, in the first dose fraction most of the cells in the sensitive region are killed. Those survive are mostly the ones in the radioresistant S phase of the cell cycle. Hence, a synchrony occurs as a result of the first dose fraction.

Then, the survival decreases and reaches a dip at about 6 hours, and again increases. The variation seen in the cell survival beyond 2 hours interval is due to progression of cells into other phases (G2, M) of the cell cycle, which is radiosensitive. This is termed as reassortment of cells. This is true only in the case of rapidly dividing cells. The dip is absent in the case of non-cycling cells. If the time interval between the split doses is greater than the cell cycle, there is an increase in survival due to repopulation of cells. Thus, the SLD repair event is summarized in three steps, namely:
- Repair of SLD
- Progression of cells in the cell cycle
- Increase in population from cell division.

The above steps are taking place simultaneously. The SLD repair basically arises from the repair of DSB. When the dose is fractionated, the damage of the first dose is repaired well before the second dose.

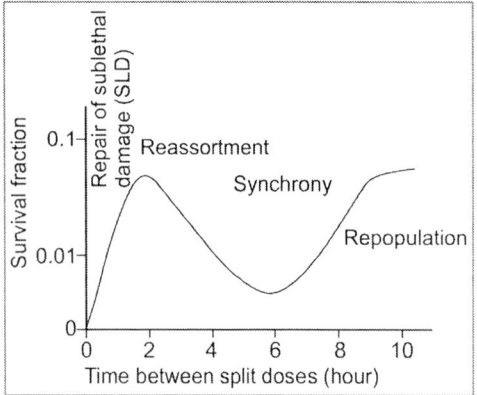

FIGURE 4.7: Survival of cells in split-dose radiation exposure separated by time interval in Chinese hamster cells.

The half-time (T½) of repair is the time duration during which 50% of the repair is completed. For mammalian cell it is about ½–1 hour for cells in culture, but longer for tissues. Full repair may take place in 6–8 hours and it can be longer for tissues. For example, central nervous system (CNS) takes more than 24 hours for full repair. The recovery ratio is a measure of SLD repair given by the survival of cells receiving a split dose divided by the survival of cells receiving the total dose as a single dose. SLD repair do occur both in tumors and normal tissue. It reflects the repair of DNA breaks before they form lethal chromosome aberrations. The repair of SLD is well correlated with shoulder of the survival curve in mammalian cells. Broad and narrow shoulder indicates the extensive and limited repair of SLD respectively. The SLD repair varies with radiation beam quality. It is significant only for low-LET radiation.

The shoulder of the survival curve is the reflection of the repair capacity of a cell line. This is nothing, but accumulation of sublethal damage to allow sublesion interactions for cell killing to occur. The lesions are repaired during split doses and resulting in the shoulder. The increase in RBE with increase of LET is the result of increased non-repairable lesions by the high-LET radiation. Repair depends on dose and time, and the maximum repair velocity is observed when damage is saturating, analogous to enzyme kinetics. The repair during irradiation is negligible at high-dose rate such as 1–5 Gy/min. This is the range of dose rate used in external beam radiotherapy and brachytherapy treatments. However, repair is significant in the dose rate range of 1.6–150 cGy/min, as practiced in LDR brachytherapy. Further reduction in dose rate will increase cell survival due to efficient repair. The dominance of repair at low-dose rate eliminates the shoulder and results in a shallower line or straight line. The survival curve gets well separated between different types of cells due to the difference in repair capacity.

Non-lethal Damage

Though non-lethal damage (NLD) hypothesis exists, it is not proved by experiments. It is not correlated with decrease of survival fraction. There may be some molecular damage, but there is no damage to the organism. This may be used to propose models in radiation protection.

BIBLIOGRAPHY

1. Charles A Kelsey, Philip H Heintz, Daniel J Sandoval, et al. Radiation biology of medical imaging. Hoboken, New Jersey: Wiley Blackwell; 2014.
2. Eric J Hall, Amato J Giaccia. Radiobiology for the Radiologist, 7th edition. Philadelphia, USA: Wolters Kluwer/Lippincott Williams & Wilkins; 2012.
3. Gorden Steel G. Basic Clinical Radiobiology, 3rd edition. London: Arnold; 2002.
4. Orton CG. Fractionation: Radiobiologic principle and clinical practice. In: Faiz M Khan (Ed). Treatment Planning in Radiation Oncology, 2nd edition. Philadelphia, USA: Lippincott Williams & Wilkins; 2007.
5. Orton CG. Radiobiology. In: Subir Nag (Ed). High Dose Rate Brachytherapy: A Textbook. Armonk, NY, USA: Futura Publishing Company Inc; 1994.
6. Rao BS. Interaction of radiation on cells. In: One Year Postgraduate Course in Hospital Physics and Radiological Physics. Mumbai: Bhabha Atomic Research Centre; 1984.

Modification of Cell Survival

CHAPTER 5

MODIFYING AGENTS

Response of cells to radiations depends upon several factors, namely the radiation quality, dose rate, dose fractionation, oxygen tension, cell cycle stage, presence of chemical protectors and sensitizers, recovery and repair process, etc. It is found that densely ionizing radiations such as alpha (α) particles and neutrons are much more effective in cell killing than low-linear energy transfer (LET) radiations such as X- and gamma (γ)-rays. A dose of radiation either delivered over a protracted period or in multiple fractions are less effective compared to that of acute irradiation. The cells in synthesis (S) phase are generally more radioresistant and cells in late gap 2 (G2) and mitotic (M) phases are sensitive to radiation. Oxygen is the best known sensitizer of radiation damage with a dose modification factor of about 3. Several other chemicals are also found capable of giving either sensitizing or protecting effects. Variation of radiosensitivity among the strains of same species is correlated to their differential repair capacity. Difference in the ability to accumulate and repair radiation damage is also the cause, which decides radiosensitivity.

Modification of radiation response occur mostly by two important mechanisms. They are:
1. The magnitude of the damage induced by radiation is reduced/increased with radioprotectors/radiosensitizers.
2. By interfering with postirradiation repair processes.

The first process occurs within 10^{-6} s following irradiation and the second process occurs several minutes after irradiation. The knowledge of modifiers of radiation damage is important for the evaluation of radiation risk and also for the improvement of radiotherapy. The various modifying agents are:
- Oxygen
- Linear energy transfer
- Cell cycle stage
- Cellular repair process
- Radioprotectors
- Radiosensitizers
- Hyperthermia
- Dose rate and dose fractionation.

The following paragraphs will explain the role of above parameters on cell survival, except dose fractionation and dose rate, which will be discussed in Chapter 8 and 9 respectively.

OXYGEN EFFECT

The oxygen effect was first observed by Swartz at Germany in 1912. Oxygen is one of the best known sensitizers of radiation damage. In the absence of oxygen (anoxic), the cells show remarkable resistance to cell killing. Hence, oxygen is a dose-modifying agent and the effect is represented by a term oxygen-enhancement ratio (OER). The OER is the ratio of the radiation dose in the absence of oxygen to the radiation dose in the presence of oxygen that produces the same biological effect:

$$\text{OER} = \frac{\text{Dose to produce certain biological effect without oxygen (hypoxia)}}{\text{Dose to produce the same effect with oxygen (oxic)}}$$

--- (1)

The OER value is always equal or greater than 1. It remains almost same value over a dose range and it is 2.5–3.5 for low-LET radiations such as X- and γ-rays (Fig. 5.1). A survival fraction to reduce to 0.01, requires a dose of 28 Gy in hypoxic and 10 Gy in oxic conditions, which results in an OER of 2.8. For a daily radiotherapy fraction of 2 Gy, the OER is around 2. For rapidly growing cells cultured in vitro, the OER is about 2.5 at low doses. It is also found that OER varies with phase of the cell cycle. Cells in G1 phase have lower OER than in S phase, since G1 cells are comparatively more sensitive to radiation. Hence, OER of asynchronous population is slightly smaller at low doses than high doses.

Oxygen effect is maximally effective when oxygen is present during irradiation. However, it is proved to be effective even if it is added after the irradiation, provided the delay is not too long (< 5 min). Experiments reveal that oxygen work at the level of free radicals. It is known from our earlier studies that radiation, such as X- and γ-rays, which interact with water produce ion pairs that in turn produce free radicals. The free radicals have larger life span (10^{-5} s) than ion pairs (10^{-10} s). These free radicals break chemical bond, resulting in chemical changes and biological end points. The extent of damage depends on the presence of oxygen. For example, if oxygen is present, the DNA reacts with free radical and forms organic peroxide, which is an irreversible form of target material. In other words, presence of oxygen changes the chemical composition of target material exposed to radiation. In the absence of oxygen, the above reaction cannot take place, the ionized target molecules are able to repair themselves and retain their function.

The mechanism of oxygen effect can be understood as follows. Oxygen interacts with hydrogen radical (H°) and form hydroperoxy radical ($HO_2°$), which persists for relatively long periods. Then, two hydroperoxy radicals react and form hydrogen peroxide (H_2O_2) and oxygen molecule (O_2). The hydroperoxy radical and hydrogen peroxide give toxicity to the biological system. The oxygen molecule repeats the cycle with hydrogen molecule as well as with organic free radicals, which are difficult to repair. This is very important, if the organic free radical happen to be a DNA molecule. Thus, oxygen produces permanent damages in the irradiated cells.

Oxygen Concentration

However, the OER increases with oxygen tension very sharply in the range of 0–20 mm Hg and reach a saturation value, corresponding to oxygen tension of 30 mm Hg (Fig. 5.2). This corresponds to 5% oxygen concentration. As the oxygen concentration increases, the cells become more sensitive to radiation. There is a rapid change of radiosensitivity as the pressure of oxygen changes from 0–30 mm Hg. A further increase of oxygen to the level of air or pure oxygen at high pressure

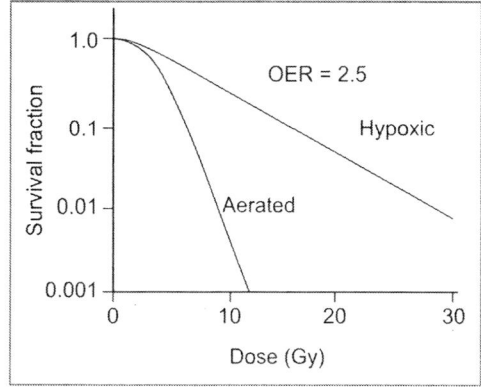

FIGURE 5.1: Change of survival curve shape due to hypoxic and aerated conditions of cells for low-LET radiations.

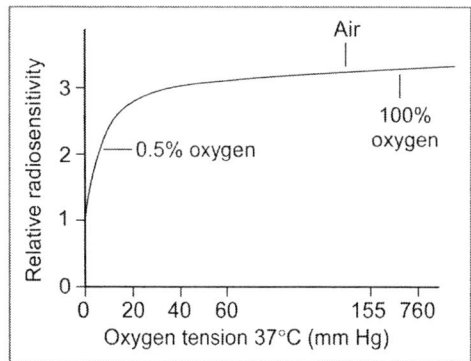

FIGURE 5.2: Relation between radiosensitivity of cells with oxygen tension

has little impact on radiosensitivity. The 50% radiosensitivity (halfway) occurs between anoxia and full oxygenation at 3 mm Hg, corresponds to 0.5% oxygen concentration.

Further, the amount of oxygen required for subjecting the cells, which are radiobiologically oxygenated, is very small with X-rays. Most of the normal tissues in the human body have the oxygen tension of 20–40 mm Hg, similar to that of venous blood, which are said to be well-oxygenated radiobiologically. However, oxygen probe measurements indicate that the oxygen tension may vary from 1 to 100 mm Hg for different tissues. This suggests that many tissues are having hypoxic cells, which has impact in radiotherapy. If oxygen pressure levels decrease below 10 mm Hg, tissues are considered hypoxic and show increasing radioresistance.

Oxygen Effect in Tumors

Histopathological studies reveal that the presence of hypoxic cells in solid tumors. There are two types of hypoxia in human body cells, namely:
- Chronic hypoxia
- Acute hypoxia.

In general, tumor does not have smooth muscle and have incomplete endothelial lining and basement membrane. Tumor cells are supplied with different oxygen concentration ranging from maximum near the capillaries and minimum away from the capillaries.

Chronic Hypoxia

Chronic hypoxia arises from the limited diffusion distance of oxygen through tissue that is respiring. The diffusion distance depends on the rapid rate of oxygen metabolism in the cells. This type of hypoxia is found in tumor cells and the cells may remain in chronic hypoxia state for long period. In 1955, Thomlinson and Gray observed first the oxygen effect in radiotherapy. According to them, tumor areas have necrotic cells surrounded by tumor cells, which appear as rings. Secondly, tumor cells can proliferate and grow only if they are very close to oxygen or nutrient supply. No necrosis was observed by them in small tumor cords with radius 160 μm. Every tumor cords with radius more than 200 μm will have a necrotic center. It is stated that the diffusion limitation of nutrients and oxygen leads to necrosis of the cells far off from the capillaries. They have estimated the oxygen diffusion distance in respiring cells as 150 μm. However, recent studies have estimated the oxygen diffusion distance as 70 μm in respiring tissue at the arterial end of a capillary (Fig. 5.3). In the case of venous end, the diffusion distance is found to be less than 70 μm.

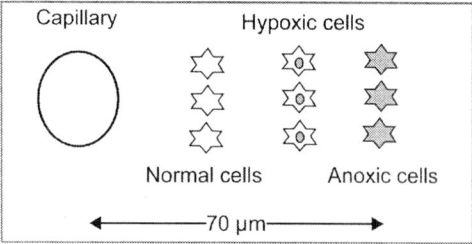

FIGURE 5.3: Oxygen diffusion status in tumor from a capillary. Normoxic cells are well oxygenated; oxygen is depleted in anoxic cells causing necrosis and hypoxic cells lie between normoxic and anoxic cells.

From the capillary, the oxygen is well supplied to the tumor up to some distance and these cells are said to be normoxic cells. At a greater distance, the oxygen supply is depleted and the cells become necrotic. The above two types of cells can be seen histopathologically as follows:
- Cells appear to be proliferating well
- Cells those are dead.

In between the normoxic and necrotic (anoxic) cells, there exist a layer of hypoxic cells. This is the region where the oxygen supply is sufficient for cells' viability and not sufficient to maintain their radiosensitivity. Hence, these cells are not killed, instead protected during radiotherapy. In this situation, the treatment fails and gives space for tumor regrowth.

Acute Hypoxia

Acute hypoxia arises in tumor cells due to temporary closure or blockage of blood vessels. This results from malformed vasculature with transient fluctuations. If the blockage is permanent, the cells die altogether. It is found that tumor blood vessel open and close in a random fashion, making different tumor regions to intermittent hypoxic state. If radiation is delivered in multiple fractions, the hypoxic cells present in the first fraction may not be there in the second fraction. Instead, another hypoxic region is present during the second fraction. In contrast, it is rare for the chronic hypoxia cells, to get reoxygenated.

There are enough evidences that human tumor contain hypoxic cells. This is proved by histologic study, oxygen probe measurements, hypoxia markers, comet assay, positron emission tomography (PET) and single-photon emission computerized tomography (SPECT) studies. The oxygen probe measurement is a gold standard, which implant electrodes directly into tumors and measure the oxygen concentration by a polarographic technique.

It gives oxygen profile of tumor with a help of computer and it is a faster technique.

Impact of Oxygen Effect in Radiotherapy

Animal experiments have shown that there are hypoxic cells in tumor and the proportion of hypoxic cells is same before and after fractionated radiation treatment. In general, tumor contains both aerated and hypoxic cells. During radiotherapy, more aerated cells are killed, since they are radiosensitive. Hence, the proportion of hypoxic cells should increase after radiotherapy, instead it remains the same. This suggests that after first fraction of radiotherapy some of the hypoxic cells become oxygenated and become sensitive to radiation. This behavior is called reoxygenation in which hypoxic cells are oxygenated after a radiation dose. These oxygenated cells are selectively killed in the subsequent fraction of radiation. That is why, the percentage of hypoxic cells remains the same instead of increasing.

The oxygen status of the cells in a tumor is not static and it is constantly changing. Reoxygenation may be slow and fast in tumors and it takes 1 hour to several days after irradiation. After radiation the tumor shrinks in size, the cells that are beyond the range of oxygen diffusion range comes closer and get oxygenated. This is a slow process that involves chronic hypoxic cells. It takes several days for tumor to shrink and get reoxygenated. The reoxygenation of acute hypoxic cells is fast that take place within hours. This is due to temporary closure of blood vessel, which is reopened quickly after radiation exposure.

The above said argument is true for human tumor also, with few variations. It is found that the radiation fraction schedule of 60 Gy in 30 fractions is sufficient enough to cure majority of tumors. This suggests that human tumor do have hypoxic cells and they get reoxygenated during fractionated radiotherapy. However, there are some tumors that do not reoxygenate quickly and it is the cause for treatment failure; radiotherapy is ineffective in that tumors. Therefore, the time scale of reoxygenation is required for tumors, to achieve effective cure in radiotherapy. Unfortunately, it is available only for few tumors and not for all tumors. Hypoxia can also decrease the efficiency of some of the chemotherapy drugs. This may be due to low pH, blood flow fluctuation, drug diffusion distance, decreased free radical production, decreased proliferation, etc.

Presence of hypoxic cells plays a vital role in tumor progression and in metastasis. Clinical studies have shown that if the oxygen level is less than 10 mm Hg, it helps tumor progression, suggesting there is a correlation between level of oxygen and aggressiveness of tumor. Even a small fraction of hypoxic cells can be a limiting factor in the treatment of tumors in radiotherapy. Several techniques such as dose fractionation, hyperbaric oxygen, high-LET radiations, hypoxic sensitizers and hyperthermia have been used to overcome the problem of hypoxic cells in radiotherapy.

RELATIVE BIOLOGICAL EFFECTIVENESS

All radiations are capable of producing same type of biologic effects, but the magnitude of the effect per unit dose differs. In other words, different radiations of equal dose do not produce the same level of biological response. For example, 100 cGy of neutron can produce more biological effect than 100 cGy of γ-rays. To evaluate the effectiveness of different radiations the term relative biological effectiveness (RBE) was introduced and it is defined by the National Bureau of Standards in 1954 as follows:

$$RBE = \frac{\text{Dose of 250 kVp X-rays required to produce certain biological effect}}{\text{Dose of reference radiation required to produce the same effect}} \quad ----(2)$$

The effects or endpoints include chromosomal mutation, cataract formation, acute lethality, etc. Historically 250 kVp X-rays were used as standard, but nowadays more than 1 MeV cobalt (Co)-60 γ-rays is used as standard. It is defined as the ratio of the dose of γ-rays (D_γ) to the dose of reference radiation (D_r) required to produce an equal amount of particular biological effect. The biological endpoint for mammalian cells is survival fraction, which is of the order of 0.1–0.01. In the case of animals, some measured functional endpoint is used as biological effect. 1 Gy of α-particle can produce much more biological endpoint than a 1 Gy γ-ray. Survival curves of human kidney cells irradiated with different radiation modalities suggests variety of biological effectiveness for a given dose. The 4 MeV α-curve is found to be linear and the 250 kVp X-ray curve appear with wide shoulder.

The RBE depends upon the cell system, radiation quality (LET), radiation dose, number of fractions, dose rate, the nature and condition of the biological system (physiological state, temperature, oxygen concentration, pressure and absence of free radicals). The LET is a parameter that describes the average energy deposition per unit path length of the incident radiation and it is expressed in keV/μm. X-rays with specific ionization of 100 ion pairs per micron of water is taken as unity. This is equivalent to LET of 3.5 keV per micron of water. RBE is higher at low doses, due to lesser efficacy of the reference radiation per unit dose at low vs high doses, i.e. low-LET radiation appears with wider shoulder.

The RBE increases with increasing LET of a particular radiation and attain maximum at about 200 keV. After that, it decreases with increase of LET, especially in mammalian cells. This is called overkill effect, where ionizations get wasted with high ionization densities. The RBE is higher for lower dose rate of low-LET radiation, but dose rate effect is absent for high-LET radiation. The RBE is lower for high dose single fraction and larger for multiple small fractions. There is also difference in RBE between tumor and normal tissues. RBE is higher for late reacting normal tissues, since they have greater repair capacity than early reacting tissues. RBE has no unit and it is an index of radiation quality for biological damage.

LINEAR ENERGY TRANSFER

The LET of a radiation is the average energy transferred per unit path length of the track in the medium and its unit is keV/μm. It is calculated as track average, by dividing the track into equal lengths and calculating the energy deposited in each length, and finding the average. Different radiations interact with medium with different mechanisms (Table 5.1). The X- and γ-rays give rise to fast electrons, which have unit electrical charge, small mass and are sparingly ionizing. Neutrons interact with protons and other heavier nuclei and give rise to recoil protons, which give further ionizations. The mass of neutron is 2,000 times higher than that of electron. The α-particles and accelerated charged particles are heavy, highly charged, slow moving and produce very dense ionizations along their path. The α-particles have two electrical charges and it is four times heavy as neutron and 8,000 times as that of electron. Hence, the spatial distribution of ionizing events produced by different particles varies widely. More densely ionizing radiation deposit larger amount of energy along their track as compared to sparingly ionizing radiation. The X- and γ-rays are called low-LET radiation, neutron as intermediate LET and α as high-LET radiation (Fig. 5.4).

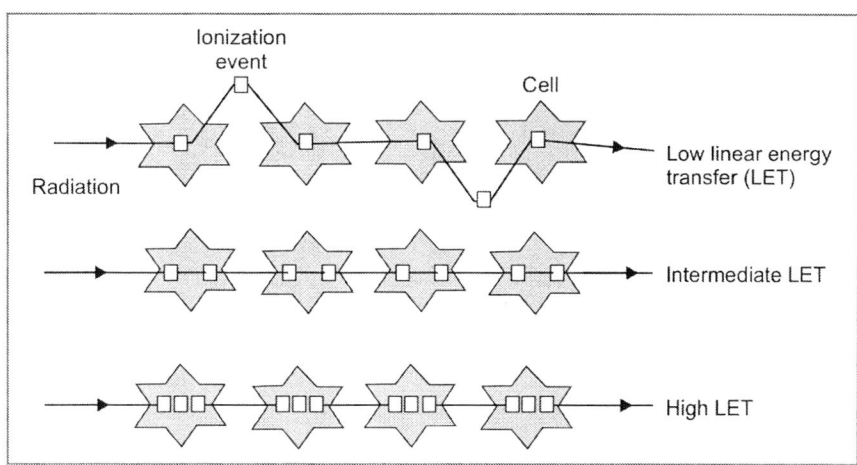

FIGURE 5.4: Distribution of ionization events in cells by the radiation. Low-LET radiations produce less number of ionization per cell. As LET increases, more ionization is produced within one cell.

Table 5.1: Linear energy transfer (LET) values for different radiations

Radiation	LET (keV/μm)
Cobalt-60 γ-rays	0.2
250 kV X-rays	2.0
10 MeV protons	4.7
150 MeV protons	0.5
14 MeV neutrons	12
2.5 MeV α-particles	166

LET-RBE Relation

Biological effects or endpoints include chromosomal mutation, cataract formation, acute lethality, etc. Generally, RBE is related to LET of the radiation as shown in Figure 5.5. For most of the biological endpoints studied RBE increases gradually with LET and reaches a maximum at about 100 keV/μm and then decreases. The RBE is proportional to LET at the beginning, which is relevant for low-LET radiations, e.g. X-rays, γ-rays and electron. Later, it increases with LET, suggesting deposition of higher energy in the tissue, which is applicable to high-LET radiations, e.g. α-rays. Beyond 200 keV/μm, the RBE decreases with increasing LET, due to overkill or wastage of radiation. This means that radiation deposits excess energy than that is necessary to kill the cell.

Mammalian cell requires more than one hit to inactivate the cell. In case of low-LET radiations such as X- and γ-rays, more than one electron should deposit energy in the cell in order to inactivate the cell. As the LET increases, more ionization is produced in the cell and that is equal to the optimum number required to inactivate the cell. If the LET increases to higher value, large number of ionizations occurs within the cell. Thus, the energy deposited within the cell is excess than that required for cell killing and the energy is said to be wasted.

For about 200 keV/μm, the radiation produced by separation of ionization events, overlaps the diameter (2 nm) of the DNA double helix. Hence, the probability of causing double strand break is more at this energy, and hence, the curve reaches the maximum. In contrast, at low-LET, the ionization events separation exceeds the diameter of the DNA, and hence, less effective. At high-LET, i.e. 200 keV/μm, the ionization events are closer and exceed within the DNA diameter and getting wasted, thus making high-LET ineffective.

The RBE-LET relation does not remain the same for all biological systems and endpoints. Apart from LET, RBE also varies with radiation quality, radiation dose, dose rate, fractionation, biological system and endpoint scored or level of effect for which the RBE is calculated, e.g. lethality, division delay, chromosome aberrations, mutation induction, etc. Radiation quality includes both radiation type and energy. Dose rate influences low-LET radiations more than the high-LET radiation, since variation of dose rate changes the shape of the survival curve more in low LET. RBE values are high for tissues that accumulate damage or capable of repairing sublethal damage.

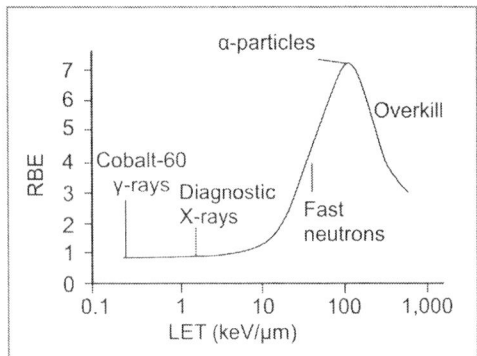

FIGURE 5.5: The relation between linear energy transfer (LET) and relative biological effectiveness (RBE) of ionizing radiation. RBE increases with LET of the radiation and reaches a maximum at 100 keV/mm and then decreases due to overkill.

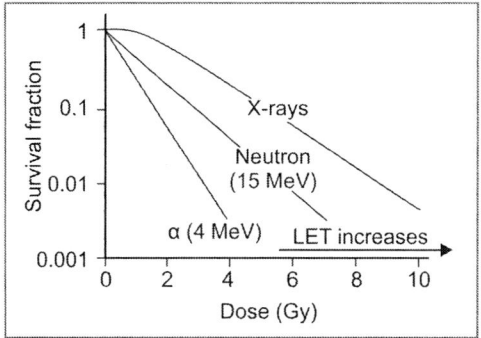

FIGURE 5.6: Survival curve shape changes with linear energy transfer (LET) of the radiation. As the LET of radiation increases, the slope of the survival curve gets steeper and size of the initial shoulder become smaller.

Variation of RBE at different survival levels occurs due to the differences in the shape of the survival curves for high- and low-LET radiations. The survival curve for high-LET radiations is a straight line or totally exponential, whereas it has large initial variable shoulder for low-LET radiations (Fig. 5.6). This means that low-LET radiation accumulates the damage or capable of repairing sublethal damage at high survival level. Hence, low-LET radiations are less effective at higher survival levels or at lower doses compared to neutrons and alpha particles. The RBE varies with survival level and it increases with decreasing dose. At low survival level, usually at high doses, the RBE is less. At higher survival level, usually at low doses, the RBE is higher. If the survival level is 0.6 and 0.01, then the RBE for neutron is 3 and 1.5 respectively. The shape of the survival curve varies with cell line, and hence, the RBE of each cell line is different.

Importance of LET-RBE in Radiotherapy

Effect of fractionation

The RBE varies with fractionation of dose and it increases with fraction. There is a difference in the survival curve between a single dose and the same dose given in multiple fractions (Figs 5.7A and B). A neutron dose delivered in three fractions will have high-RBE compared to the same neutron dose delivered in a single fraction. As we fractionate more, the dose per fraction becomes small and the RBE is larger. In addition, fractionation causes multiple shoulders and the width of the shoulder refers to the wastage of dose. The shoulder

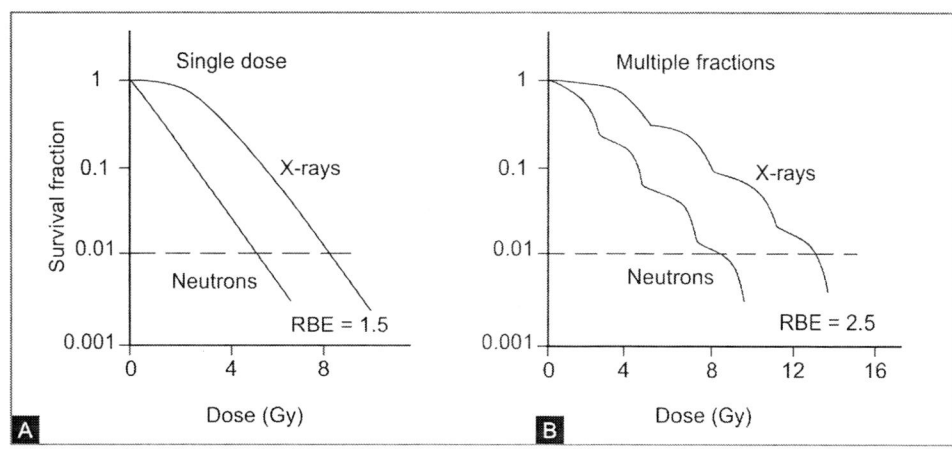

FIGURES 5.7A and B: Survival curves of X-rays and neutron. **A.** Single dose; **B.** Fractionated dose of multiple fractions. The relative biological effectiveness (RBE) increases with fractionation for a given survival level.

width varies between X-rays and neutron. The neutron shoulder width is lesser than that of X-ray, hence wastage of dose is less in neutron. Thus, neutron is more effective with more fractionation or small dose per fraction, than low-LET X-rays. This is also true for continuous low-dose rate radiation, compared to an acute exposure. Hence, low-energy neutrons are more effective than X-rays in radiotherapy.

Effect of OER and RBE as a function of LET
The OER value decreases with increase in LET of radiation; and for α-particle, it approaches to one (Fig. 5.8A). The OER is about 3 for low-LET radiation; it decreases from 30 keV/μm onwards and reaches unity above 200 keV/μm. Thus, the OER is 1.6 for 15 MeV neutron and 1.3 for 4 MeV α-particle. That means the densely ionizing radiation produces high amount of damage in its track and the survival curve does not have an initial shoulder. In the case of fast neutrons with intermediate LET, the OER is about 1.6. The survival curve has much reduced shoulder. Hence, oxygen effect is large and very important, only for low-LET radiations such as X- and γ-rays. It is absent for densely ionizing radiations such as α-rays. On the other hand, OER decreases with increase of RBE (Fig. 5.8B). It is found that the raise of RBE and rapid fall of RBE occurs at about 200 keV/μm.

CELL CYCLE STAGE

The percentage of cell survival varies with phases of the cell cycle. The cell lines exposed to radiation show a marked difference in survival for different phases of the cell cycle. When population of synchronized cells are exposed to a fixed dose of radiation, the level of survival observed varies by a factor of nearly 40 depending upon the position of the cells in the mitotic cycle. The most sensitive cells are those in M and late G2, and hence, their survival curves are steeper and have no shoulder (Fig. 5.9). The cells in the late S phase are radioresistant and the survival curve is less steep with broader shoulder. The cells in the G1 and early S phase have intermediate sensitivity. Variation of cell sensitivity with the cell stages also depends upon the cell line.

The open structure of DNA may explain the radioresistance in G1 phase. The reason for resistance in S phase is the greater availability of undamaged sister template, which increases the homologous recombination. Additionally, there may be conformational changes in DNA, which facilitates repair process. Greater amount of homologous recombination (HR) repair than by non-homologous end joining (NHEJ) repair takes place in late S phase, which is responsible for its resistance. On the other hand, there is little time

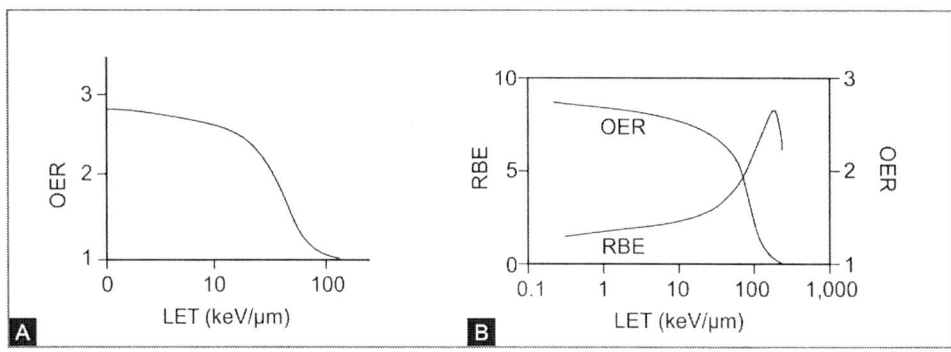

FIGURES 5.8A and B: Relation between OER, LET and RBE. **A.** Variation of oxygen-enhancement ratio (OER) with linear energy transfer (LET); **B.** Variation of OER and relative biological effectiveness (RBE) with the function of LET.

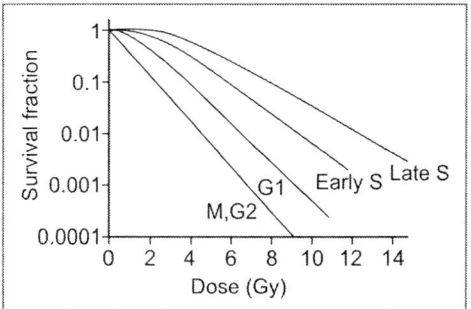

FIGURE 5.9: Survival curves for various phases of the cell cycle shows the variation of radiosensitivity, suggesting that the cells in the M and G2 phases are most sensitive to radiation and curve has no shoulder.

for cell in G2 phase to repair radiation damage before going for division. The chromatin compaction and poor repair competence due to reduced enzymes access are the reason for high radiosensitivity of G2/M phase. Hence, the surviving cell population becomes synchronous, immediately after radiation.

The OER varies with different phases of the cell cycle. For a given phase of cycle, the OER remains same for all dose levels. However, in a asynchronous population, the OER varies as a function of dose. The OER is smaller at higher level of survival, at which the cells are most sensitive (M). The OER is larger at higher doses and lower level of survival, at which the cells are more resistant (S). The S phase exhibits largest OER in the cell cycle. Though the concept is interesting, it has little relevance in radiotherapy.

Causes of Variation in Radiation Sensitivity at Different Phases of Cell Cycle

The causes for the variation of sensitivity in cell cycle may be due to:
1. Variation in the physical form of nuclear material such as DNA during S phase.
2. Variation in the repair capacity of the cells at different stages of the cell cycle.

The variations in sensitivity at different phases of cell may have implications in radiotherapy. When a fraction of dose is delivered to the tumor, the cells become synchronized as a result of G2 block. The position of the cells in the cycle during the second fraction may play an important role in deciding the degree of killing. Furthermore, it is likely that the cells can be treated with drugs, which arrest the cells in a specific phase of the cell cycle that may be very sensitive to the action of another drug. Though it is proved in cell culture, it is difficult to apply in human radiation treatments.

RADIATION PROTECTORS

Chemical compounds provide protection against radiation if present during radiation. They cause vasoconstriction through metabolism and reduce oxygen concentration in organs. Since cells are less sensitive under hypoxia, the chemical offers protection. Such radioprotective substance includes sodium cyanide, carbon monoxide, epinephrine, histamine and serotonin. The well-known radioprotector is the sulfhydryl (SH) group compounds that include cysteine (Fig. 5.10A) and cysteamine (Fig. 5.10B). The ratio of the radiation dose required to bring about a certain level of effect in the presence of the drug to that required in its absence is referred to as dose reduction factor (DRF):

$$DRF = \frac{\text{Dose of radiation to produce certain effect with drug}}{\text{Dose of radiation to produce the same effect without drug}} \quad \text{-----(3)}$$

The maximum DRF that can be obtained is about 1.8. The survival curve gets altered with drug and exhibit broad shoulder (Fig. 5.11). The DRF values can be increased up to 2.5–3, the value close to the oxygen-enhancement ratio.

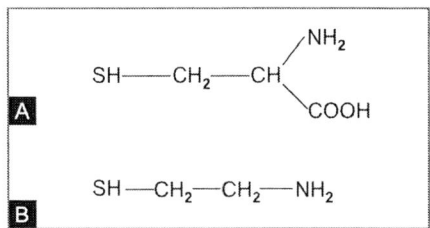

FIGURES 5.10A and B: Structure of radioprotectors. **A.** Cysteine; **B.** Cysteamine.

Radioprotectors act mainly by modifying the indirect effects of radiation caused by free radicals. The SH group mechanism includes:
- Free radical scavenging
- Hydrogen atom donation.

In general, radiation produces free radicals and in the presence of oxygen, free radicals form peroxy radicals, which are biologically very reactive. The SH prevents such oxygen-based free radical generation by radiation as well as by chemotherapy. They also donate hydrogen atom to facilitate direct chemical repair at sites of DNA damage and restore the structure of DNA to its original state. In addition, the protectors can interact with radiosensitive groups of the target molecule and protect it from the action of radiation. Radioprotectors offer much less protection against high-LET radiation as compared to X-rays. The reason for this is that most of the killing induced by high-LET radiation occurs by direct effect and the amount of local damage is greater. Sulfhydryl compounds are naturally available in cells and carry out cellular repair process. Some time it is added artificially to increase the radioprotection, which can be toxic to the cells. However, this toxicity can be reduced, if the SH group is covered by a phosphate group, e.g. thiophosphate.

Radioprotectors in Radiotherapy

Addition of cysteine or cysteamine is protective with a DRF of 1.8 in a whole-body-irradiated mice. Though the cysteine and cysteamine are effective radioprotectors, they are too toxic and cause nausea and vomiting. Hence, attempts were made to find new drugs with reduced toxicity. They are basically drug involving SH group covered by a phosphate group, namely:
- Cystaphos (WR-638)
- Amifostine (WR-2721).

Amifostine is sold in the market as Ethyol and is approved by Food and Drug Administration (FDA), USA. It is a phosphorothioate compound, basically a prodrug that is activated by dephosphorylation in patient. The alkaline phosphatase enzyme present in the normal tissues and capillaries, dephosphorylates amifostine and converts it into an active metabolite WR-1065. Then, the metabolite penetrates into the normal cells and scavenges free radicals generated by ionizing radiation. It is found that it reaches the normal tissues rapidly and acts slowly in tumors, because of hypoxic and chaotic vasculature. That is why, radiation is given immediately after the drug is injected. Thus, the drug will manifest a differential effect between normal and tumor tissues during radiation. It is used for the prevention of xerostomia in patients with head and neck cancer, without affecting the local control. The drug is administered daily 30 minutes before each dose of radiation, since the drug penetrates at slower rate in tumors. This will reduce symptoms such as

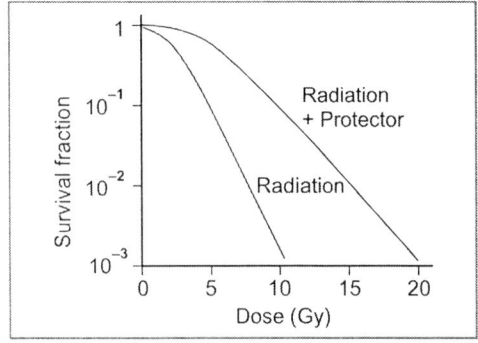

FIGURE 5.11: Survival response of mammalian cells to gamma (γ)-rays in the presence and absence of radioprotector.

dry mouth, difficulty in eating and speaking. Amifostine is a phosphorothioate, nonreactive and does not penetrate cells. Though it is protecting normal tissue in head and neck squamous cell carcinoma, but its role in tumor is still controversial. Whether it is offering protection to the tumor or not, still not very clear in radiotherapy.

Amifostine is useful as protector both in radiotherapy and chemotherapy. It reduces nephrotoxicity, ototoxicity and neuropathy from cisplatin and hematologic toxicity from cyclophosphamide. The drug is effective with a dose of 25 mg/kg and the optimal dose for cytoprotection is 400 mg/kg, which is toxic with side effects. It has also some protective effect on tumors; hence, its role in clinical radiotherapy is limited. Amifostine is also used as protector against mutation and carcinogenesis. Sodium selenite, pentoxifylline and vitamin E all show clinical benefits in reducing morbidity, e.g. less xerostomia, mucositis, proctitis, enteritis and fibrosis.

RADIATION SENSITIZERS

Several chemical and physical agents sensitize cells to the action of radiation. Oxygen is the best known sensitizers with a dose modification of about 3. The sensitization is a free radical process and the sensitizers should mimic oxygen. They enhance radiation damage by fixing damage produced by free radicals. It is believed that the sensitizers are not rapidly metabolized in the tumor cells. They can penetrate further and could reach the hypoxic cells and make them sensitive. Thus, hypoxic cell radiosensitizers increase the sensitivity of hypoxic cells, not the aerated cells. The efficiency of sensitization is related to electron affinity of the compounds. The magnitude of sensitizing effect is expressed by the sensitizer enhancement ratio (SER). The SER is the ratio of the radiation dose without sensitizer and radiation dose with sensitizer.

$$SER = \frac{\text{Radiation dose without sensitizer}}{\text{Radiation dose with sensitizer}} \quad \text{-----(4)}$$

The SER value is more than 2, which is obtained in animal experiments. The different sensitizers are grouped into three categories. The first group consists of DNA base analogous, which when incorporated into DNA renders it sensitive to radiation, e.g. 5-bromouridine. They enhance the production of SSB and DSB by a factor of 2-3, upon radiation. The second group is compounds, which enhance the radiation damage through the formation of active transients. Electron-affinic agents and iodine-containing compounds come under this category. The electron-affinic compounds have greater affinity for electrons and acts similar to chemical replacement for molecular oxygen. They sensitize the hypoxic cells, which are believed to be present in the human cancer. These compounds may interfere with the cation-anion recombination, which restitute the damage. The third category involves agents, which modify the biochemical processes. Inhibitors of energy metabolism, DNA synthesis and repair pathways fall into this category. Each sensitizer seems to act by one or more of the above mechanisms.

Sensitizers in Radiotherapy

Study of hypoxic sensitizers has great implications in radiotherapy of tumors containing sizeable fraction of hypoxic cells. Chemical sensitizers not only substitute for oxygen, but penetrate into tumors more easily than oxygen. This is the reason why hypoxic sensitizers are far more effective than hyperbaric oxygen in vivo.

The requirement of sensitizers include:
1. It should selectively sensitize hypoxic cells.

2. Chemically stable.
3. Highly soluble in water.
4. Effective at low doses.

The optimal drugs having the above characteristics are misonidazole, etanidazole and nimorazole. Misonidazole and etanidazole are 2-nitroimidazoles and nimorazole is a 5-nitromidazole. Though the misonidazole has been tested in animals successfully, its clinical trial is not encouraging. The maximum dose of misonidazole that can be given during radiotherapy treatment is 0.5 g/m^2. The etanidazole is equally efficient as misonidazole, has short half-life, with lesser toxicity. It has low neurotoxicity, penetrates poorly into nerve tissue and does not cross blood-brain barrier. The clinical result of the drug is found to be little beneficial to the patients. The nimorazole is less effective radiosensitizer, but also less toxic compared to misonidazole and etanidazole. Hence, large doses can be given and it is tested and proved as effective in head and neck cancer in clinical trial.

There are other molecules that increase the radiosensitivity of cells using clonogenic assays including molecules that enhance DNA damage. They are halogenated pyrimidines, inhibitors of DNA repair, modifiers of cell cycle checkpoints such as caffeine and modifiers of mitogen-activated protein kinase signaling pathways such as inhibitors of Ras, epidermal growth factor receptor (EGFR) or protein kinase B (AKT). Their application in clinic requires some clarity on specificity for tumor cells vs normal cells. Recent studies have been focusing on molecular differences such as levels of gene expression or mutations in critical genes such as protein 53 *(p53)*. Differential uptake of halogenated pyrimidines into DNA in place of thymidine in proliferating cells provides one rationale. Both bromodeoxyuridine (BrdU) and iododeoxyuridine (IrdU) have been studied clinically.

These molecules are slightly larger than thymidine and partially disrupt the structure of the DNA making it more susceptible to radiation damage. As on date, these molecules are not in clinical use, due to the difficulty in obtaining enough differential uptake between tumor and normal tissue. Inhibitors of epidermal growth factor receptor (EGFR) have recently been tested in the clinic with some success. EGFR is highly expressed on some tumor types, e.g. head and neck squamous cell carcinoma and non-small cell lung cancer, and inhibition of the signaling pathway is believed to reduce proliferation of tumor cells and block stimulation of this pathway by the radiation treatment, but the exact mechanisms of the effect remain uncertain. The Cetuximab, an anti-EGFR molecule is approved for concurrent chemoradiation therapy in recurrent/metastatic head and neck squamous cell carcinoma.

Other approaches include antisense oligonucleotides to inhibit the expression of antiapoptotic factors such as Bcl-2; gene-directed enzyme prodrug therapy targeting DNA synthesis; radiation-activated molecular switches to drive specific promoters in tumors to increase expression of toxic molecules such as tumor necrosis factor-alpha (TNF-α); inhibitors of checkpoint kinases (Chk1 and Chk2) required for expression of cell cycle blocks conceivably because blocking abrogation would inhibit repair.

HYPERTHERMIA

Mammalian cells subjected to heat treatment at temperatures above 43°C show appreciable sensitization to X- and γ-rays. Magnitude of sensitization increases with temperature duration in the temperature range of 41–45°C. It is believed that hyperthermia has additive and synergistic radiosensitizing properties in the above temperature range. Heating tumors in situ with temperature range of 39–42°C can activate vascular, metabolic and immunologic parameters of the tumor environment. This may further bring heat-induced cell killing in tumor and radio- or chemo-sensitization of cells.

The survival curves with heat are similar in shape of radiation survival curves (Fig. 5.12). However, the curve shapes with high and low temperature are significantly different. Especially, the curves are flattened out at low temperatures. Hence, the mechanism of cell killing is different in both radiation and heat. The major difference between heat and radiation is the protein damage. Application of radiation and heat gives additive effects. If radiation is given alone, the cells die, while attempting the next mitotic division. In the case of heat and radiation, the damage happens early due to necrotic and apoptotic mechanisms. In addition, heat:
- Affects differentiating and dividing cells
- Induces chromosome aberrations in S phase cells
- Affects DNA repair process.

The heat also makes the hypoxic cells sensitive to heat. Thus, hyperthermia revert the hypoxic nature of cells and improve vascular perfusion. The inherent sensitivity of tumor and normal cells do not vary much towards heat. Further, the damage caused by heat is acute in nature unlike the late effects of radiation.

Thermal Enhancement Ratio

Thermal enhancement ratio (TER) is the ratio of the X-ray energy required to produce a given level of biological damage with radiation alone to that for combination of radiation and heat. The typical value ranges from 2 to 4 for animal experiments, but it is between 1.15 and 1.5 for human clinical study. However, there is a possibility of increasing the TER value in human treatments.

It is possible to enhance cell killing with moderate heat that can produce an acceptable damage to normal tissues. Solid tumors are characterized with poorly vascularized parts with deprived nutritional status and increased acidity. Heat exploits the above parameters and enhance cell killing in tumor volume. Thus, tumor is at higher temperature than normal tissue due to poor vasculature in tumors. To minimize thermal damage to normal tissue and improve energy coupled during hyperthermia, heat is delivered to the target tissue using water as a coupling medium. Temperature controlled water cools the skin or superfacial normal tissues during hyperthermia.

Once tissue is heated, it becomes resistant to further heating. This type of thermal-induced resistance is called thermotolerance. The onset of thermotolerance depends on the initial temperature typically defined for temperatures greater than 43°C. Thermotolerance in surviving cells occurs few hours after heat and exists for few days. Hence, hyperthermia fractions were initially designed with a gap of few days with limited heat fractions, twice per week. Subsequently, it was demonstrated that inhibition of DNA damage repair can happen at lower temperatures, as low as 40°C. Thus, combining more hyperthermia fractions is likely to enhance heat-induced radiosensitization for every radiation fraction.

Unlike animal studies conducted in temperature-controlled water bath, the thermal dose delivered using heating devices in patients is not really uniform due to hot and cold regions in tumor tissue. This causes

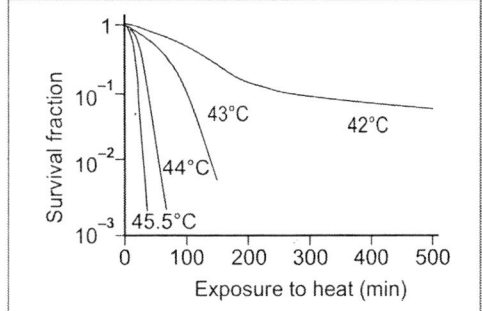

FIGURE 5.12: Survival curves for mammalian cells heated at varying temperatures and length of time

difficulty in establishing temperature-time relationship. Hence, a normalized time-temperature method called cumulative equivalent minute (CEM) concept at 43°C was introduced by Sapareto and Dewey. This is a thermal isoeffect dose formulation and is given by:

$$\text{CEM, } 43°C = tR^{(43-T)} \quad \text{----- (5)}$$

where,

t is the time of treatment
T is the mean temperature during heating
R is a constant = 0.25.

This metric converts temperature history during the treatment into equivalent number of minutes at 43°C. This has been widely adopted to report thermal dose delivered in several randomized clinical trials with slight modifications in the temperature distribution (T), such as minimum, median or percentile of the temperature distribution. Overall, higher CEM indicates better treatment outcome.

Heating Methods

Most of the in vitro studies employ hot water baths, warm air incubators and immersion of animals in warm water, etc. as methods of heating. However, in human, microwave (MW), radiofrequency (RF)-induced currents and ultrasound are employed to produce heat. Microwaves give good localization at shallow depths, while RF has deeper penetration and lacks spatial localization. In the case of ultrasound, the presence of bone or air cavity distorts the heating, but gives uniform distribution in soft tissues.

Hyperthermia can be used for tumor ablation too. Thermal ablation involves temperatures of the range of 50-60°C, which can kill largest number of tumor cells in shortest time. Ablative heating is done with the help of RF or microwave applicators, inserted into the tumors, under the guidance of CT or ultrasound. As the heat is given, a tissue coagulation zone develops around the applicator. The blood ceases inside the zone that changes vascular perfusion, oxygen state, pH and metabolism of the tumor environment. The coagulation zone size is about 3-5 cm diameter depending on the number of needles used in the treatment.

Hyperthermia Equipment and Safety

The hyperthermia equipment is basically divided in two groups, namely:
- Local and regional (Fig. 5.13)
- Whole hyperthermia.

The power generator is ultrasound, RF or MW source, whose power can be controlled either manually or automatically. It has impedance matching network, power meters and indicators. The power is transferred to applicator via tuning and matching mechanism. The applicator consists of piezoelectric transducer, waveguide or an antenna to deliver power over a given surface of tissue. The power density distribution depends on the applicator design and specific absorption rate (SAR). It is desirable to cool the skin surface.

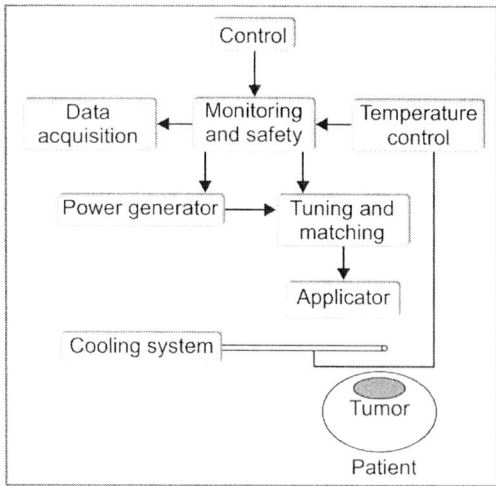

FIGURE 5.13: A typical hyperthermia system for local and regional application

The cooling system consists of air blower or cooling pad through which precooled fluid is circulated. The cooling system is integrated with bolus or the applicator. Bolus is an interface material that lies in between the applicator and patient body surface. It also facilitates efficient transfer of power to the body.

The temperature system consists of temperature sensors, leads to carry signals, electronics parts and calibration setup. Computer-controlled software provides feedback to the power controller and records temperature-time curves for a number of temperature probes. The computer for power control and data acquisition monitors thermometer reading, feedback signals, and controls power delivered during treatment. All the above are displayed on a monitor screen and also recorded.

Any hyperthermia equipment must be tested for its performance before its clinical use. The American Association of Physicists in Medicine (AAPM) and Radiation Therapy Oncology Group (RTOG) have given the guidelines for such performance testing.

Clinical Applications

Hyperthermia found wide range of application in cervical cancer, breast cancer, metastatic malignant melanoma, metastatic nodes, head and neck cancer, glioma and esophageal cancer. Though lot of trials has been conducted, its regular use in clinical set up is yet to be established.

BIBLIOGRAPHY

1. Charles A Kelsey, Philip H Heintz, Daniel J Sandoval, et al. Radiation Biology of Medical Imaging. Hoboken, New Jersey: Wiley Blackwell; 2014.
2. Eric J Hall, Amato J Giaccia. Radiobiology for the Radiologist, 7th edition. Philadelphia, USA: Wolters Kluwer/Lippincott Williams & Wilkins; 2012.
3. Jacob Van Dyk. The Modern Technology of Radiation Oncology, Vol. 2. Madison, Wisconsin: Medical Physics Publishing; 1999.
4. Rao BS. Interaction of radiation on cells. In: One Year Postgraduate Course in Hospital Physics and Radiological Physics. Mumbai: Bhabha Atomic Research Centre, 1984.

Biological Effects of Radiation

RADIATION EFFECTS

Radiation exposure can destroy cell's ability to reproduce, but may not disturb its function. Then, there is a chance that cell may not complete the cell cycle and undergo reproductive death. These cells do not reproduce, but continue its function such as production of protein, hormones, etc. The reproductive death is not related with cancer. In general, the biological damage depends upon the mitotic index of the cell. This means that cell with higher replication rate and high turnover, are more sensitive and exhibits their damage immediately. The organ response depends upon the mixture of cells by which it is made up of. Biological effect of radiation is generally divided into stochastic and deterministic effects (Figs 6.1A and B).

Stochastic Effect

A stochastic effect is one in which the probability of effect occurring increases with dose rather than its severity, e.g. induction of cancer. A radiation dose of 5 Gy has more probability to induce cancer than 1 Gy dose, even though the severity is same in both cases. It has no threshold dose; even a small radiation has the ability to cause stochastic effect. The risk increases with increase of dose and there is no dose at which the risk is 0. Stochastic effects are the principle health risk from low-level radiation; hence, it is important in medical exposures of patients and occupational workers.

With respect to stochastic effect the dose-effect relationship is studied only in a group of human population. The dose-effect relationship is linear, but its effect < 100 mSv (low-dose range) is not verified. The observed dose response at higher doses is extrapolated for low doses. This is applicable for radiation protection and safety purpose. There is no method to identify the appearance of effect in an individual. The increase in occurrence of effect can be proved only by epidemiological studies. The stochastic effects are further classified as:
- Radiation carcinogenesis
- Hereditary effects.

The stochastic effects usually appear as late effects (Fig. 6.2).

Deterministic Effect

A deterministic effect is one in which the severity of the effect increases with dose due

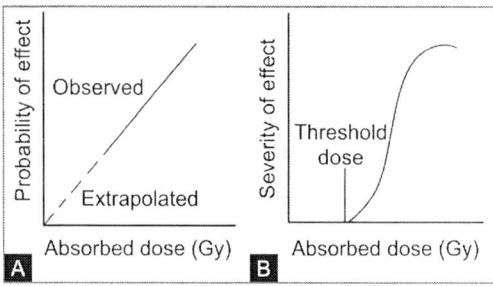

FIGURES 6.1A and B: Biological effects of radiation. A. Stochastic effect; B. Deterministic effect of radiation.

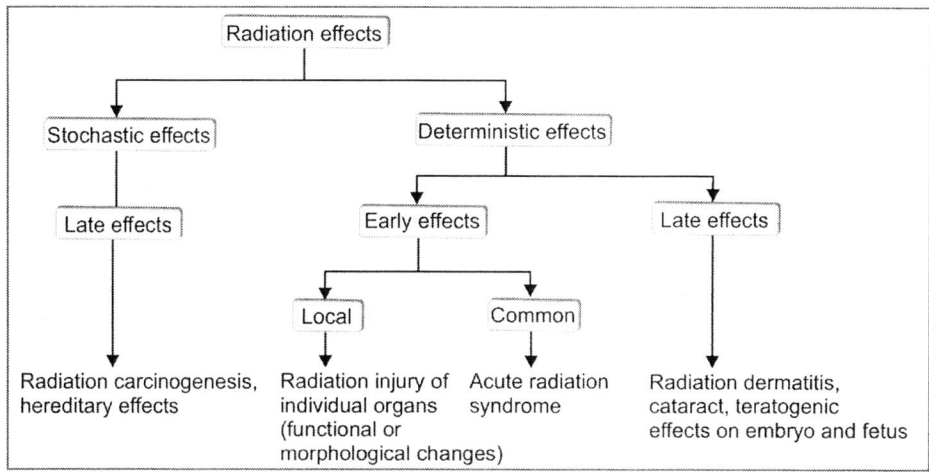

FIGURE 6.2: Various types of radiation effects after acute and chronic radiation dose

to degenerative changes in tissues. It has a threshold dose below which no radiation effect is seen. Deterministic effects always occur at high radiation doses. The threshold dose is higher than the doses from natural radiation or from the occupational exposure at normal operations. Skin erythema, cataract and hematopoietic damages fall under this category. Deterministic effects are often known as normal tissue reactions, which are given in detail in Chapter 7 'Normal Tissue Reactions of Radiation'.

There is a time interval between irradiation and the occurrence of radiation effects called latent period (Fig. 6.3). It is the indicator of tissue radiosensitivity and it is inversely proportional to the dose. This means that lower the dose, greater the latent period and appearance of the effect. There are two latent periods correspondingly for the appearance of early and late effects. Of course, there is no difference in the radiation damage in terms of early and late effects, but only difference is the time of appearance of effect. The acute early effects appear within a month after radiation exposure. The early effects are further divided into:

1. Radiation injury in individual organs or normal tissue reactions.
2. Acute radiation syndrome.

The former is the result of partial body exposure including radiotherapy treatments and latter is the outcome of whole-body exposure due to accidents. The late effects appear within few months or year after radiation exposure that includes:
- Radiation dermatitis
- Cataract
- Teratogenic effects on embryo and fetus.

The basic difference between stochastic and deterministic effect arises from the interaction. In the former, the radiation interacts with DNA at the cellular level.

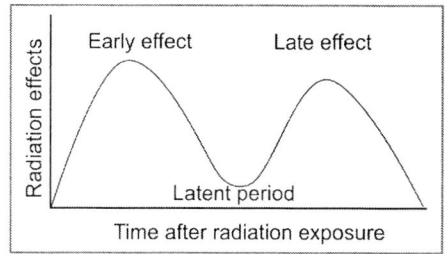

FIGURE 6.3: Biomodel plot of radiation response over a period of time after exposure, early and late effects with latent period

This results in mutation, which can be passed to the next generation, leading to genetic effects or induction of cancer. In the latter, the interaction is in a group of cells that forms the tissue or organs. This will end up with loss of tissue function.

RADIATION CARCINOGENESIS

Radiation can induce cancer in the form of delayed somatic effect. Radiation is weak carcinogen at low doses, especially at medical exposures. The radiation-induced carcinogenesis is a stochastic effect in which the probability of occurrence increases with dose, without threshold dose. Radiation that is sufficiently given to any part of the body increases the incidence of malignant tumor. Thomas Alva Edison's assistant Clarence Dally died of cancer caused by diagnostic X-ray exposure. Marie Curie and her daughter Irene died due to leukemia as a result of radiation exposure. The human experience comes from survivors of Hiroshima and Nagasaki, the patients who were exposed to medical radiation and other occupations such as:

1. Atomic bomb blasts on Hiroshima and Nagasaki—about 120,000 have been followed and about 50,000 received dose > 5 mSv. Solid tumors and leukemia were reported.
2. About 14,000 patients were given radiotherapy in spine for ankylosing spondylitis and for pain relief, resulted in leukemia (Britain, 1935–1944).
3. Thyroid cancer in children irradiated for benign conditions (enlarged thymus). Since thyroid was included in the treatment field, both benign and malignant tumors including breast cancer were reported.
4. Chernobyl accident released large amount of radioiodine that increased the incidence of thyroid cancer in children. In addition, they took cow's milk, which is concentrated with radioiodine, since the cow eats contaminated foliage.
5. Female patients treated with X-rays for postpartum mastitis and other benign conditions received 1–6 Gy and reported incidence of breast cancer.
6. Female patients in the Nova Scotia Sanatorium had repeated fluoroscopy for pulmonary tuberculosis, which landed in increased incidence of breast cancer, since the patient dose is about 0.8–0.9 Gy.
7. Uranium miners were exposed to radon gas in a closed environment. It deposited on the bronchial epithelium along with radon decay. Their lung is exposed to intense alpha (α) rays, resulting in excess lung cancer.
8. Girls engaged in USA in painting watch and clock dials with luminous paint containing radium developed excess cancer of bone. Radium was injected when they used to put the brush in their mouth for making it pointed. Since radium is a bone seeker, bones used to be irradiated to large doses.
9. Patients who had radium-224 injection for the treatment of tuberculosis or ankylosing spondylitis ended with excess bone cancer.

Radiation-induced cancer cannot be differentiated from other cancer caused by other agents. The probability of radiation-induced cancer depends on:
- Radiation quality
- Dose rate and fractionation
- Latent period, age and gender.

Radiation-induced tumors usually have a long latent period and it depends upon dose. The latent period varies with the type of cancer induction and leukemia has the shortest latent period. The national and international scientific bodies have come out with conclusions that:

1. The cancer risk varies with cancer site.
2. Risk is higher at young age.
3. Risk is greater for females than males.
4. Solid cancer risk obeys linear function of dose.

5. Leukemia follows non-linear function of dose.
6. Generally cancer risk is elevated over 50+ age for all cancers, except leukemia.
7. No evidence for radiation-induced genetic effects.

Epidemiological Studies on Radiation-induced Cancer

The study of atomic bomb survivors involves 44,771 deceased members. Among this, there are 9,335 cancer deaths including 582 deaths from leukemia. Analysis of this study reveals that about 440 cancer death and about 100 deaths from leukemia can be related to radiation exposure. This comes around 4% and 15% respectively. Significant relationship is found in stomach, colon, lung, leukemia, breast, esophagus, bladder, ovary and liver, for radiation-induced cancer. It is also found that radiation have not induced some of the common cancers that is existing in population. This includes prostate, cervix and rectum. However, it is not possible to make generalized statement for radiation-induced cancer and death.

Another prospective study known as life span study (LSS) involves about 120,000 people, including 90,000 people who are reconstructed for radiation dose. Among this, 5,000 people are projected to receive a whole-body dose of > 1 Gy. This study have found that persons exposed to radiation as children or adult, develop radiation-induced cancer after a long latent period than the persons exposed in their later life. This means that radiation-induced cancer tends to occur at the time when age-related spontaneous cancer risk occurs. This statement is based on relative risk model, which state that irradiation causes a dose-dependent increase of relative risk of developing certain types of cancer.

Factors Affecting the Cancer Induction

Radiation Quality

Radiation quality influences the cancer induction. Since RBE varies with LET, high-LET radiation can cause more damage to DNA than low-LET radiations. The damage induced by the low-LET radiations is easily repaired, whereas it is unlikely in high-LET radiation. Hence, the mutation induced by the high-LET radiation is much more than low-LET radiation. Mostly these damages are irreparable and there is higher probability of cancer induction. More cancer incidence has been reported in human lung, thyroid and bone by the high-LET radiation, for a given dose, relative to low-LET radiation. It has been reported that low-energy photons are more effective in producing dicentrics in lymphocytes than high-energy photons. This may be due to low-energy secondary electrons suggesting higher RBE for low-energy photons and electrons. However, epidemiological studies do not support this concept.

Dose Rate and Fractionation

The radiation-induced damage decreases in cells at low doses. This is due to the ability of the cells to repair their damage at low doses or dose is fractionated. Fractionating the large doses reduces the incidence of carcinogenesis especially in leukemia. Hence, the dose and dose rate effectiveness factor (DDREF) are used to convert the high-dose risk estimate into low-dose risk estimate. It is defined as the factor by which radiation cancer risks observed after large acute doses should be reduced when radiation is delivered at low-dose rates. The International Commission on Radiological Protection (ICRP) recommended a DDREF of 2, whereas Biological Effect of Ionizing Radiation (BEIR) VII reports suggested a factor of 1.5. As per BEIR, if the dose is spread out over a period of time at low-dose rate, then if it is delivered in an acute exposure, the incidence of malignancies is lesser. This suggests that there is dose rate effect even at low-dose rate radiation. However, the dose rate used in the medical diagnostic and therapeutic procedure does not influence the cancer risk.

Latent Period, Age and Gender

The latent period for induction of cancer vary with age and gender. The minimum latent period is 2–3 years for leukemia and it is proportional to age at the time of exposure. The corresponding latent period for solid tumors is 5–40 years. The induction of ovarian cancer is three times higher for a 10-year-old female than 50 years old. In the case of whole-body exposure, females have 40% higher risk for radiogenic cancer than males. Thus, women have higher risk for breast, ovarian and lung cancer; whereas, it is colon for men. However, the risk for liver cancer is low in women than men.

BEIR V and BEIR VII Reports

The National Research Council Committee on the Biological Effects of Ionizing Radiation (BEIR) Report V has been published in 1990 on the title 'Health Effects of Exposure to Low Levels of Ionizing Radiation'. As per the report, the radiation-induced mortality at low exposure level is 4% per Sv and it is 8% for high-dose rates. This is in agreement with the ICRP estimate, which is 4.2% per Sv for a population of adult workers and 5.7% per Sv for the whole population (includes young ones). The BIER V Committee advocate the linear dose-response model for all cancers except for leukemia and bone cancer for which linear quadratic model is recommended.

The BEIR VII Report (2006) gives the most up-to-date and comprehensive risk estimates for cancer and other health effects from the exposure of low dose, low-LET radiation in humans. It provides risk estimates based on scientific evidence and epidemiology. The report assumes the low dose as 100 mGy and chronic exposure as dose rate less than 0.1 mGy per min, irrespective of the total dose.

It provides risk estimates for both cancer incidence and cancer mortality for a population similar to US. It also provides organ-specific risk estimates adjusted to gender, age at exposure and time interval after exposure. It also emphasizes heritable effect and risk from in utero radiation exposure. It supports a 'linear-non-threshold' (LNT) risk model that the risk of cancer proceeds in a linear fashion at lower doses without a threshold and that the smallest dose has the potential to cause a small increase in risk to humans. There is a linear dose-response relationship between exposure to ionizing radiation and the development of solid cancers in humans. It is unlikely that there is a threshold below which cancers are not induced; but at low doses, the number of radiation-induced cancers will be small.

Other health effects (such as heart disease and stroke) occur at higher radiation doses, but additional data must be gathered before an assessment of any possible dose response that can be made between low doses of radiation and non-cancer health effects. The report also concludes that with low dose or chronic exposures to low-LET irradiation, the risk of adverse heritable health effects to children conceived after their parents have been exposed is very small compared to baseline frequencies of genetic diseases in the population.

Dose-response Models

Scientists have developed dose-response models to predict the risk of cancer in human population due to low-level radiation exposure. These models led to dose-response curves, whose shape are LNT, linear quadratic (LQ) and linear threshold (Fig. 6.4). The LQ dose-response curve suggests reduced incidence of cancer at low doses and higher incidences at higher doses. In this, the probability of the effect is proportional to dose (linear) and square of the dose (quadratic). They represent the response to low dose, high dose and high-dose rate. The degree of curvature refers to the ratio of quadratic and linear coefficients (βD^2 and αD). These coefficients can be estimated for different types of cancer in order to predict the dose-response curve.

In general, the radiation-induced effects are lesser at low dose, since the repair mechanism is active in cells. This is also true in the case of radiation treatments delivered in multiple fractions. LNT curve is extrapolated backwards from high doses and dose rates. If it is corrected for dose rate effect with DDREF, it will predict lesser incidence of cancer. The BEIR Report VII has used the DDREF of 1.5 to correct the observed high-dose response to low-dose response. This is represented by the back projected dotted lines in the LNT model. There is an uncertainty in the estimation of DDREF due to inconsistency in the animal data. The DDREF is also sensitive to the selection of dose range that is used for estimation. BEIR VII Committee used DDREF corrected LQ dose response for estimating cancer risk for all cancers except leukemia.

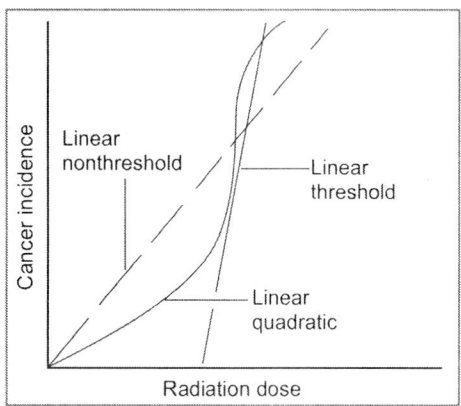

FIGURE 6.4: Dose-response models for radiation

The linear threshold curve predicts that there is threshold for cancer induction. This is true in some cancers such as sarcoma, rectum, etc. whereas untrue in prostate, pancreas, etc. Hence, the linear threshold model cannot be accepted as a universal model. Therefore LNT hypothesis is considered as the suitable model until the uncertainties including the DDREF is cleared. It overestimates the incidence of cancer at lower doses from low-LET radiation. Hence, it is used at present in radiation protection for estimating risk. Considering the radiation levels used in medicine (< 100 mSv), the difference between the LNT and LQ model dose response is very small.

The risk estimates were revised in 1990 after reassessing the Japanese atomic bomb survivors, which resulted in increase in risk estimation. The reasons for the above are:

1. Revised neutron dosimetry, which incorporated greater risk from low-LET radiation.
2. Large number of cancer cases seen among bomb survivors.
3. Revised model projects risk beyond the period of observation.

This is because that the bombs used in the two cities are different. The bomb used in Nagasaki is type tested one, emitted gamma (γ)-rays with few neutrons and its dosimetry was based on partly measurements. On the other hand, the bomb used in Hiroshima is not type tested, emitted both γ-rays and neutrons, and its dosimetry was based on computer simulations.

Risk and Risk Models

Radiation risk is a probability that a given individual will incur a deleterious effect as a result of a dose of radiation. The BEIR-VII and ICRP-103 both have expressed risk as:
- Relative risk
- Absolute risk.

Relative Risk

Relative risk (RR) is one way of expressing the risk from radiation in an exposed population. It states that irradiation causes a dose-dependent increase of relative risk of developing certain types of cancer. The overall risk obtained by his/her spontaneous cancer risk multiplied with dose-dependent risk factor. The spontaneous risk increases with age. In other words, it is the ratio of the cancer incidence in the exposed population to that in general population. For example, a relative risk of 1.2 predicts a 20% (120/100) increase over the spontaneous rate of cancer incidence:

Excess relative risk (ERR)
= 1.2 − 1
= 0.2 = 20%

To detect the above relative risk (1.2), with a statistical confidence of 95% ($p < 0.5$), when the spontaneous incidence in the population is 2%, one must require a study sample of population of > 10,000. Thus, RR provides information about the magnitude of the increased risk, over the natural incidence.

The ICRP estimated the life time risk as 10% after an acute radiation exposure of about 1 Gy. If it is delivered over a period of weeks or months, the risk is reduced to about 5%. If it is spread over a working life, it is about 4%.

Absolute Risk

Absolute risk is another way of expressing risk. It is expressed as number of excess radiation-induced cancers in 10^4 population per Sv per year. The excess absolute risk (EAR) is the difference between two absolute risks and expressed as EAR per unit dose. This is mostly used in radiation epidemiology.

Example 1: The risk is 4 per 10,000 person-Sv and the latency period is 10 years. What is the risk of developing cancer in the next 30 years, from a dose of 0.1 Sv?

Actual duration = 30 − 10 = 20 years
Dose = 0.1 Sv
Risk = 4×10^{-4}
Risk of developing cancer
= $20 \times 0.1 \times 4 \times 10^{-4}$
= 8×10^{-4}

If 10,000 people are exposed to a dose of 0.1 Sv, 8 additional cases of cancer will be seen in that population in the next 30 years.

Multiplicative and Additive Model

There are two types of risk models, namely:
- Multiplicative model
- Additive risk model.

Multiplicative model: It is a relative risk model, which assumes that the excess cancer risk increases in proportion to the baseline cancer rate. It predicts that after a latent period, the excess risk is a multiple of the natural age-specific risk for the specific cancer for a given population. It accounts for age and cancer risk at the time of exposure. The risk increases with age (predicts greater risk at older age). The Figure 6.5A describes the differences in magnitude of the projected cancer risk for a given exposure at different ages.

Additive risk model or absolute risk model: It predicts a fixed increase in risk, unrelated to the spontaneous population and age-specific natural cancer risk at the time of exposure (Fig. 6.5B). In this model, a constant increment in incidence is added to the spontaneous disease incidence throughout the life. Both models do not describe cancer risk adequately, after exposure. However, the multiplicative model is consistent with scientific evidence for radiation-induced cancer. The scientific evidence says that:

1. Radiation acts as initiator of carcinogenesis, rather than as promoter.
2. Children carry more radiation risk than adults.

Hence, this model is used by the BEIR VII Committee, for the estimation of tissue-specific solid cancer risk as a function of age and

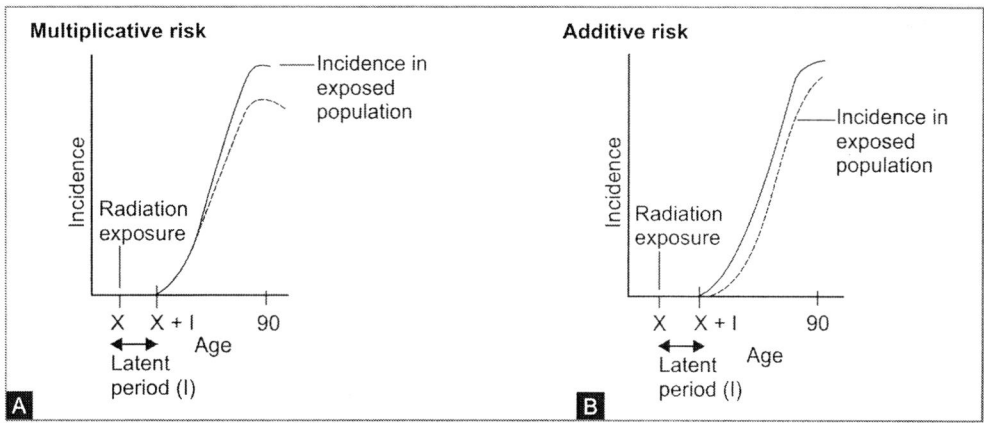

FIGURES 6.5A and B: Risk models. A. Multiplicative model; B. Additive model (*Note:* X is the time of age at exposure and l is the latent period).

gender. However, both the above models can be used to transport risk calculated from one population to another dissimilar population. The BEIR VII adopted multiplicative model in some cancer (thyroid) and absolute model in some cancer (breast), as hybrid policy.

Risk Estimate for Population

The ICRP risk estimates for radiation-induced cancer and genetic effect (detriment) for low-dose radiation is 4.2% per Sv for adult population and 5.7% per Sv for whole population. Detriment is the total harm to health experienced by an exposed group and its descendants, as a result of the group's exposure to a radiation source. The corresponding ICRP estimate for cancer induction is 4.1% and 5.5% respectively for adult and general population. The BEIR VII Committee estimate of radiation-induced cancer incidence is 13.7% and 9.0%, per Sv, for female and male respectively. The corresponding mortality rate is 6.6% and 4.8%, per Sv, for female and male respectively. The BEIR VII Committee have also analyzed the atomic bomb survivors and stated that cancer incidence is a function of age at the time of exposure for male and females (Fig. 6.6). The risk decreases with increase of age, and children and adults are vulnerable for radiation-induced cancer. The ICRP estimate of 5.5% per Sv for cancer risk of general population is very close to the BEIR VII estimate of 5.7% per Sv.

The lifetime probability of natural incidence of cancer in the US population is 41%, including both female and males. Assuming the LNT model, if 100 people are exposed to 100 mSv, the additional cancer cases is about 1 over the lifetime of the population group. However, it is difficult to identify that person in the population who has additional risk of cancer caused by the radiation. The above projection of 100 people exposed to 100 mSv at a time is not possible in practice, and hence, not realistic. The maximum

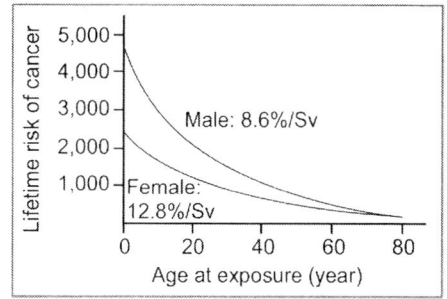

FIGURE 6.6: Cancer incidence with age for males and females—100,000 persons exposed to a single dose of 0.1 Gy from atomic bomb data (BEIR VII).

radioactive material, which may expose a 10,000 population with a dose of 10 mSv in one year. The additional cancer risk with the above radiation is found to be 0.3% over the natural frequency.

Radiation-induced Cancers

Leukemia

Leukemia is a rare disease in the general population, but radiation-induced leukemia is the most common. Leukemia may be acute or chronic with lymphocyte and myeloid form and it is the first malignancy related to radiation. Acute and chronic myeloid leukemia is mostly reported in radiation exposure. The survivors of atomic bomb blasts of Hiroshima and Nagasaki, and patients treated for ankylosing spondylitis for pain relief with X-rays are the human evidences. Leukemia can be detected within 1-2 years after exposure and it reaches a peak after 12 years, and influenced by the age at the time of exposure. The incidence of radiation-induced leukemia decreases with increase of age (Fig. 6.7). Younger age group is more vulnerable with shorter latent period. The BEIR VII uses LQ model for risk assessment for leukemia as a function of age at exposure. In this approach, there is no need for DDREF correction, since it uses LQ model. However, it is more appropriate to express the risk as a function of time. The BEIR VII risk estimate for leukemia is 70 and 100, for females and males respectively for an exposure of 0.1 Gy per 100,000 persons, with an age distribution similar to US population.

Thyroid Cancer

Thyroid is very high sensitivity organ for radiation-induced cancer, especially in children, it is less effective in adults. Thyroid cancers are mostly well-differentiated papillary adenocarcinoma with lower percentage of follicular form and grow slowly.

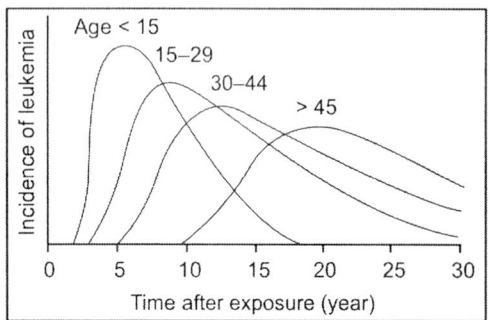

FIGURE 6.7: Incidence of leukemia with age at the time of exposure from atom bomb survivors

This can be removed surgically or treated with radioactive iodine, hence the mortality rate is low. There is a wide difference in the incidence of spontaneous and radiation-induced thyroid cancer. There is also gender difference, as women have two to three times greater risk than men. The incidence of thyroid cancer is about 2% of the annual total cancer incidence. However, its mortality rate is very low (0.2%). Its absolute risk is lower than that of breast, colon and lung.

The dose response for thyroid cancer follows a linear curve. The latent period is about 5-35 years for benign nodules and about 10-35 years for thyroid malignancies. The internal radionuclides are equally inducing thyroid cancer as external radiation like X-rays. The NCRP (2008) report has predicted that I-131 and I-125 are 30-60% effective as external radiation, in causing thyroid cancer. It also suggested that I-123 and Tc-99m is also effective as external radiation in causing thyroid cancer. Irradiation also causes hypothyroidism and thyroiditis. The threshold values are 2 Gy for external radiation and 50 Gy for internal irradiation. However, children have lower threshold values.

Iodine mostly concentrates in thyroid gland, and hence, the thyroid dose is about 1,000 times higher than that of other organs of the body, which is a major concern in children. Most of the data come from Japanese survivors, residents of Marshall island, individuals who took radioactive iodine after Chernobyl accident, children treated with X-rays for cancer, tinea capitis and enlarged thymus. Their dietary deficiency might have enhanced their thyroid cancer occurrence. The observed thyroid cancer is a papillary carcinoma, which is common in adults and children.

Breast Cancer

Breast cancer is the most common cancer in women. One out of every eight women develops breast cancer. The lifetime mortality risk for breast cancer is about 2.8% with variation with race and ethnicity. The other factors that enhance breast cancer are age at the time of first pregnancy, family history and estrogen levels in blood. Women with no children or one child have greater risk for breast cancer. Reduced risk is also found to be associated with women who conceive earlier and who breastfed for longer time. Family history of breast cancer may enhance the risk two to four times. Breast cancer human data is obtained from female atom bomb survivors of Japan, patients screened with fluoroscopy repeatedly for tuberculosis, and females treated for postpartum mastitis and other benign conditions.

The BEIR VII adopted the linear dose-response model to estimate breast cancer risk, based on atomic bomb survivor data. About 800 mGy is required to double the natural incidence of breast cancer. The latent period is about 10–40 years and young women have longer latency. The BEIR VII risk estimate for breast cancer is 310 per 100,000 population exposed to a dose of 0.1 Gy. The risk is age dependent; it is 13 times higher at 5 year age, than at 50 years. Women undergoing chest CT scan with a dose of 20 mGy have a risk of 0.14%, 0.06% and 0.014% for the ages of 10, 30 and 50 respectively. Mammography is thought of a potential hazard, which will increase the incidence of breast cancer. However, the mammography-related risk is decreasing due to the advent of quality assurance and digital mammography.

Lung Cancer

Radiation is one of the carcinogenic agents in lung. The other agents are cigarette smoking, asbestos, chromium salts, mustard gas and hematite. Human data were obtained from atom bomb survivors, patients treated with ankylosing spondylitis and underground miners exposed to radon. Excess number of lung cancer was reported in the uranium miners of Colorado plateau (US), Czechoslovakia, non-uranium miners of Sweden and fluorspar miners of Newfoundland. It is difficult to separate the risk from the carcinogenic agents, since individuals are exposed to multiple agents. BEIR VI Report estimated the excess of lung cancer as 10% out of 150,000 lung cancer deaths in USA.

Bone Cancer

Radiation-induced bone cancer has only little evidence. Human data were obtained from children exposed to X-ray for tinea capitis treatment, patients treated for ankylosing spondylitis with X-rays. These numbers are very small and it is very difficult to arrive on risk estimate. The other data includes watch dial painters who have ingested radium and patients had radium-226 injections for tuberculosis and ankylosing spondylitis. These population are exposed to α-radiation and with higher radiation dose > 5 Gy. One has to extrapolate this data to lower dose levels to arrive a risk estimate, but there is a possibility of overestimate.

Skin Cancer

Radiation exposure to skin may onset neoplasms, followed by chronic dermatitis with long latent period. The possible cancers are squamous cell carcinoma and basal cell carcinoma, which can be treated effectively, if detected early. The present technology follows well-equipped safety standards, and no possibility of above such cancer incidence by radiation. There is a large difference between the incidence and mortality.

HERITABLE EFFECTS

Ionizing radiation can cause DNA damage. During the cellular process, the enzymes correct the damage and return the DNA to normal sequence and structure. Sometimes the cellular process may fail, resulting in alterations in DNA that leads to lethality or heritable changes in the surviving cells. Thus, ionizing radiation is capable of producing heritable effects. These heritable changes may be mutations or chromosomal aberrations. The heritable changes induced in the germ cell can be transmitted to the successive generations and cause disease of one kind or another. The changes induced in the somatic cells have small, but finite probability contributing to carcinogenic process.

Radiation-produced mutations are similar to that of other mutation agents, but the proportions of the different types are not the same. Studies on germ cell mutation on experimental organism and in mammalian somatic cells, suggest that all radiation-induced mutation involves changes in large segments of the DNA, such as deletion that often encompass more than one gene. As a result, radiation induces the kinds of molecular changes that can derange a genome and lead to cancer. Alternatively, if the changes occur in germ cells, become incompatible with embryo development. This may result in developmental abnormalities or lethal mutations in the germline, which would result in non-viable progeny.

The radiation-induced genetic effects were first observed in fruit fly drosophila, by Muller HJ (1927) with X-irradiation. This is followed by gamma, beta, alpha and neutron radiation, which have shown to induce both gene and chromosomal mutations. To study the heritable diseases in human, a genetic risk study with atomic bomb explosion was initiated in Japan. However, it was not informative due to heterogeneity of the exposed human population. Hence, homogeneous mice population having a spontaneous mutation rate of 1:100,000 was experimented. This mutation rate is doubled by a dose of 1 Gy, which is known as doubling dose. This may be true to human population also; however, it may not give useful information on human genetic risk. Extension of further research in the field has concluded the following:

1. Radiation can cause mutation.
2. Majority of the mutation is harmful to the organism.
3. Radiation-produced mutations are not unique.
4. Chronic radiation exposure causes few heritable effects, than acute radiation.
5. Theoretically, even a single radiation-induced mutation can cause genetic effects and it obeys LNT model.

However, no genetic effect is observed in humans, even though it is extensively studied in Japanese bomb survivors, and children of cancer survivors who underwent radiotherapy. No evidence of increased risks of malformation, neonatal death, stillbirths, chromosomal abnormality and gene changes are observed. Thus, dose associated with medical exposure causes smaller risks compared to spontaneous incidence of genetic effects.

Genetic Diseases

Genetic diseases are generally divided into:
- Mendelian
- Multifactorial.

Mendelian disease arises from mutation in single genes, whereas multifactorial disease is due to joint action of multiple genetic and environmental factors.

Mendelian

Disease caused by mutations in single gene is known as mendelian diseases. It is caused by mutations in microlesions and gross lesions. Microlesions include single base pair substitutions, deletions, insertions or duplications involving one or more base pairs. Gross lesions include whole gene or multigene deletions, complex rearrangements, and large insertions and duplications. The microlesion dominates the spectrum of mendelian diseases.

Mendelian changes the structure of DNA, which involves either base composition or the sequence, or both. The alteration is so small, which involves only substitution, gain or loss of single base. However, it causes significant heritable effects. The relation between mutation and disease is direct, simple and predictable. Mendelian diseases are further divided into:
1. Autosomal dominant.
2. Autosomal recessive.
3. X-linked, depending upon the chromosome location and transmission patterns of the mutant genes.

In a autosomal dominant disease, a single gene is sufficient to cause disease, e.g. achondroplasia, neurofibromatosis, Marfan syndrome, etc. Autosomal recessive disease require homozygosity for disease manifestation (i.e. mutation to occur in both genes obtained from father and mother), e.g. cystic fibrosis, phenylketonuria, Bloom's syndrome, etc. The X-linked recessive disease is due to mutations in genes located on the X chromosome, e.g. Duchenne's muscular dystrophy, Fabry's disease, steroid sulfatase deficiency, etc. The overall frequency of mendelian disease in the human population is about 2.4%. Mendelian diseases are rare and contribute lesser to the heritable diseases.

Multifactorial Diseases

The multifactorial diseases originate from the joint action of multiple genetic and environmental factors. The genetic basis of the multifactorial disease is the presence of a genetically susceptible individual, who may not develop the disease depending on the interaction with other genetic and environmental factors. The examples are congenital abnormalities at birth (cleft lip, neutral tube defects), adult onset diseases such as diabetes, hypertension and coronary heart diseases. The causes for the above diseases are genetic, environmental, nature of family and population. There is no simple relationship between the mutation and diseases. About 6% live births are affected by congenital abnormalities with genetic component and about 65% will develop chronic diseases at later age with genetic component. The chronic diseases include diabetes, hypertension, and coronary heart diseases. However, environmental factors play a vital role in the above incidence. Multifactorial plays a major role in causing heritable diseases.

Loss of Function and Gain of New Function

Mutation can be classified as:
1. Loss of function.
2. Gain of new function, on the basis of function.

Point mutation can abolish normal gene function, by partial or total gene deletion, disruption of the gene structure by translocations or inversions, etc. Generally, loss of function mutation in enzyme-coding genes is recessive, since half of the gene product is sufficient for normal functioning. But, loss of function mutation in genes that code for structural proteins are dominant phenotypes through haploinsufficiency or dominant negative effects. The negative effect is one in which mutant gene not only loses its function but also prevents the product of the normal allele from functioning in a heterozygous organism. On the other hand, gain of function is likely when only specific changes cause a given phenotype. Gains are usually novel functions, except cancer. But, in inherited diseases, mutant gene is expressed in wrong time in development, in the wrong tissue, in response to wrong signals or at an inappropriately high level. The spectrum of gain of function mutations mostly restricted and may not produce disease.

Epidemiological Studies

Though mutations are seen in culture, it is unable to demonstrate radiation-induced hereditary effects. The mutation rates are low in human than found in insects. The study of 27,000 individuals with screening of 28 specific proteins in atomic bomb survivors reported only two mutations caused by radiation exposure of the parents. It is also found that radiation is not increasing the sex-linked lethal mutations, which is responsible for the increased prenatal deaths of males. The irradiation of testes has increased the incidence of translocations in spermatogonical sperm cells, though additional chromosomal aberrations are not detected.

A study of 70,000 children, whose parents are likely exposed to a radiation dose of about 0.15 Gy to their gonads, was undertaken in atomic bomb survivors in Japan. No dose-dependent increase in the frequency of biochemical mutation was found in that study. It concludes that direct determination of heritable radiation effects is impossible. About 90,000 radiological technologists with 100,000 children in the US are studied for childhood cancer during 1921–1984. However, no sufficient evidence is obtained for increase of radiation-induced childhood cancer in the offspring. Similar results are obtained from the children of parents, who underwent radiotherapy treatment for cancer. The latter two studies again reconfirm the conclusion of atomic bomb survivors study. Hence, radiation-induced heritable damage cannot be determined by epidemiological studies, but based on animal experiments.

Genetic Risk

To estimate the risk for human heritable diseases, there is no foolproof information available. Hence, the mouse doubling dose is combined with available information of human population genetics, to arrive at genetic risk. For this, two information are very vital. One is the spontaneous mutation rate, and the second is the doubling dose. Then, two correction factors are very vital, the first one should allow that all mutations are not causing the disease. The second should allow the fact that the mutations used in mouse experiments are not the representative of heritable diseases of humans (ICRP-2000).

The term genetically significant dose (GSD) is used as an index to express the genetic impact of radiation-induced mutation in germ cells, in a given population. It provides an estimate of the effect of man-made radiation on the genetic pool. The GSD (Gy) is an estimated dose, which if given to the gonads of the entire population, would produce the same genetic effect as is actually observed. It is not the estimate of the genetic effects of radiation. It is the way of comparing the various sources of radiation on possible genetic effects. For example, the GSD dose is 200 µGy in USA.

Doubling dose (DD) is another term used to measure the sensitivity of the population to radiation-induced genetic damage. The doubling dose is the amount of dose required in a generation as many as mutations as those arise spontaneously. It is estimated as a ratio of the average rates of spontaneous and induced mutations in a given set of genes:

$$DD = \frac{\text{Average spontaneous mutation rate}}{\text{Average induced mutation rate}} \quad \text{-----(1)}$$

The reciprocal of the DD is the relative mutation risk (RMR) per unit dose. The genetic risk is usually obtained by combining the population genetic data in humans and the radiation genetic data in mice. The risk is the probability that an offspring of the exposed person will develop heritable diseases. The risk was estimated by using baseline disease frequency (P) or prevalence and DD:

Risk per unit dose = P × (1/DD)   ----- (2)

However, later this equation is modified by incorporating mutation component (MC). The MC is defined as the relative increase in disease frequency per unit relative increase in mutation rate:

Risk per unit dose = P × (1/DD) × MC   ----- (3)

Inclusion of MC is a unifying attempt at predicting how the frequencies of different classes of genetic diseases in the population will change, as a result of increase in the mutation rate. The spontaneous mutation rates are 5×10^{-6} per locus and $7\text{--}15 \times 10^{-4}$ per gamete chromosomal abnormalities. The human doubling dose is about 1 Gy per generation, as extrapolated from animal studies.

The BEIR VII Report suggests that an exposure of 10 mGy to the parents can cause 30–47 additional genetic disorder per 1 million births. This may be due to increase of sex-linked dominant mutations and autosomal dominant mutations. The natural background radiation does not contribute to genetic risk. Even a 100 mGy could cause only 400 additional genetic effects per 1 million live births; it is roughly 0.4% per Gy. The genetic risk can also be expressed in the following format (IAEA, 2010):

Risk = Prevalence × (1/DD) × MC × PRCF   ----- (4)

The MC refers to the relation between increased mutation rate and the rate of additional diseases, and it is less than 1. The reason for this is that majority of the existing mutations are inherited from parents and grandparents over many generations. Doubling the rate of dominant mutation will cause only 30% increase in disease in the first generations and 15% in the second generations. In the case of recessive mutations, the MC ≈ 0, since it require mutation in both alleles of the same gene. The relation between mutation and diseases is also very remote. The corresponding MC value for multifactorial diseases is ≈ 0.01. The above risk estimate is valid for protracted irradiation conditions such as radiotherapy treatment.

The PRCF stands for potential recoverability correction factor, which is ≈ 0.01. It accounts the difference in molecular structure between radiation-induced mutation and spontaneous mutation. Spontaneous mutations are generally point mutation, whereas radiation-induced mutations are large deletions, affecting the whole gene, leading to premature termination of pregnancy. The PRCF value is about 0.15–0.3. Based on the above values, one can estimate the risk < 0.1% for dominant mutation and recessive mutation. But, for the children, it is approaching to zero. Recent understanding is multisystem congenital abnormalities that include mental retardation, growth retardation and malformation. These are similar to single gene diseases, very rare and the risk for 1 Gy is about 0.1%.

The ICRP estimate of radiation-induced genetic risk is about 0.2% per Gy, taking into considerations up to two generations. The relative contribution of cancer and heritable risks to the total detriment is given in Table 6.1. The contribution of heritable risk to the total detriment is less. This is also reflected in the reduction of radiation weighting factor for gonads from 0.2 to 0.08.

Table 6.1: Normal risk coefficient for stochastic effect at low-dose rate (% per Sv) (ICRP)

Category	Cancer	Heritable	Total
All	5.5	0.2	5.7
Adult	4.1	0.1	4.2

Most of the diagnostic and occupational radiation exposures are unlikely to cause significant genetic anomalies. Delaying the conception after a radiation treatment may reduce the probability of transmission of genetic damage to the offspring. Hence, it is a recommended practice in therapeutic treatment of parents.

BIBLIOGRAPHY

1. Charles A Kelsey, Philip H Heintz, Daniel J Sandoval, et al. Radiation Biology of Medical Imaging. Hoboken, New Jersey: Wiley Blackwell; 2014.
2. Committee on the Biological Effects of Ionizing Radiation. Health Effects of Exposure to Low Levels of Ionizing Radiation. BEIR Report V. Washington, DC: National Academy Press; 1999.
3. Committee on the Biological Effects of Ionizing Radiation. Health Effects of Radon. BEIR Report VI. Washington, DC: National Academy Press;1999.
4. Committee on the Biological Effects of Ionizing Radiation. Health Risks from Exposure Low Levels of Ionizing Radiation. BEIR Report VII phase 2. Washington, DC: National Academy Press; 2006.
5. Eric J Hall, Amato J Giaccia. Radiobiology for the Radiologist, 7th edition. Philadelphia, USA: Wolters Kluwer/Lippincott Williams & Wilkins; 2012
6. International Atomic Energy Agency. Radiation Biology: A Handbook for Teachers and Students, Training Course Series 42. Vienna: International Atomic Energy Agency; 2010.
7. International Commission on Radiological Protection. Pregnancy and Medical Radiation. New York: Elsevier ICRP Publication 84; 2000.
8. International Commission on Radiological Protection. Recommendations of the International Commission on Radiological protection. ICRP Report 103. New York: Elsevier; 2007.
9. Jerrold T Bushberg, Anthony Seibert J, Edwin M Leidholdt JR, et al. The Essential Physics of Medical Imaging, 3rd edition. Philadelphia,USA: Wolters Kluwer/Lippincott Williams & Wilkins; 2012.
10. National Council on Radiation Protection and Measurements. Risk to the Thyroid From Ionizing Radiation. NCRP Report 159. Bethesda, MD: National Council on Radiation Protection and Measurements; 2008.
11. Thayalan K. Textbook of Radiological Safety. New Delhi: Jaypee Brothers Medical Publishers (P) Ltd; 2010.

Normal Tissue Reactions of Radiation

CHAPTER 7

DETERMINISTIC EFFECTS

Ionizing radiation is known to induce both stochastic and deterministic effects. In Chapter 6 'Biological Effects of Radiation', we have discussed the biological effects, especially the stochastic effects of radiation. This chapter is going to emphasize on deterministic effects with special reference to non-tumor effects. Basically deterministic effects are the one in which its severity increases with dose. It may appear as degenerative changes in the exposed tissues. This usually happens at higher doses and with a threshold dose, below which the effect do not occur, e.g. skin erythema and fibrosis. However, there is an individual variation towards radiation exposure, and hence, the threshold dose may vary for a given biological system. Radiation can cause both short term and delayed effects of varying nature among irradiated individuals. Both somatic and teratogenic effects have been observed in humans as well as in animals upon irradiation. Somatic effect is the one, which occurs on the exposed individual within his life time. Teratogenic effects are the one, which occurs on growing fetus. Thus, deterministic effects may be divided into:
1. Radiation effects on tissues or radiation injury of individual organs.
2. Acute radiation effects.
3. Radiation effects on embryo and fetus.

The consequences of acute and fractionated irradiation, and both partial and whole-body irradiation will be discussed in detail in the following paragraphs.

RADIATION EFFECTS ON TISSUES: EARLY AND LATE

Radiation treatment can leads to loss of function of the normal tissue. It may occur due to damage of mature cells or damage of stoma and vasculature. In certain cells such as bone marrow, there will be loss of proliferative capacity. The radiation-induced localized organ effects depends on radiation quality, such as linear energy transfer (LET), dose, dose rate and radiosensitivity of the cells in that organ. The effect is measured in terms of morphological and functional change of the organ. The organ exhibits its response over a period of time. If the dose is higher, shorter will be the period of expression. However, every organ has a threshold value below which no visible changes are seen. It is difficult to distinguish the radiation-induced pathology and the naturally occurring disease pathology.

However, radiation-induced effects can be healed by means of cellular regeneration and repair. Regeneration refers the replacement of damaged cells of the organ by cells of the same type, which retain the functional integrity. Repair is replacement of the damaged cells by different cell types, e.g. fibrosis. The organ is not returned to its preirradiation status; instead, its function is compromised. The organ healing depends on the dose, volume of tissue damaged, regenerative capacity and radiosensitivity of the tissues involved. Regeneration starts within a day after irradiation. If the radiation is very high, it may lead to tissue fibrosis and necrosis. The effects of radiation on tissue vary widely, in terms of radiation dose and timing of the expression of damage. The radiosensitivity of different normal tissue can be obtained by in situ assays. The mice and rat experiments reveal that the survival curves differ for different normal tissues (Fig. 7.1). Radiation effects are basically divided into early or acute and late responses or simply early and late effects. The clinical symptoms of early reaction may appear within few weeks after radiation, whereas that of late reaction takes months to years.

Early Effects on Normal Tissues

The early or acute effects occur in tissues with rapid cell renewal, in which cell division is required to maintain the tissue function of the organ. Many express their damage during mitosis and radiation can cause early death, and loss of cells. This is due to the death of large number of cells that occur within a month or few weeks of irradiation, e.g. epidermal layer of skin, oral mucosa, intestine, bone marrow, hematopoietic system and testis. These tissues

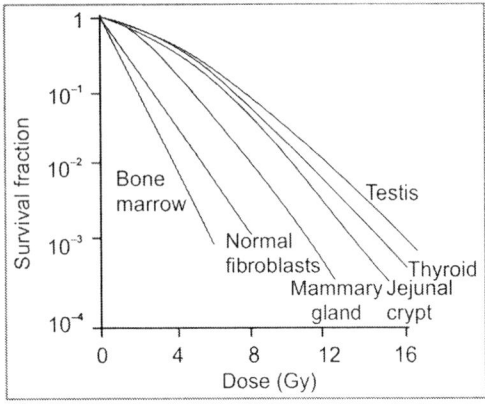

FIGURE 7.1: Survival curves for different cells of normal tissues by in situ clonogenic assays

contain epithelium cells or hematopoietic cells and connective tissue cells. The damage is the result of parenchymal components, which has short functional life span. The response is decided by the stem cells and their differentiating offspring. The time of onset of early effect correlates with life span of the mature functional cells. Early damage is repaired rapidly by the proliferation of stem cells and is completely reversible. In rare cases, it may lead to death of the individual.

Late Effects on Normal Tissues

Late reacting normal tissue shows their response over month or year after irradiation. It occurs in slow proliferating tissues, such as lung, kidney, heart, liver and spinal cord. Basically, late effects are due to the damage of connective tissue or connective cell, which arises from the loss of parenchymal cells, or blood vessels. However, the loss of parenchymal cell dominates the vascular damage in most situations. The parenchymal cells divide infrequently (liver or kidney), or rarely (central nervous system or muscles) in normal conditions. The depletion of parenchymal cell population is due to entry of cells into mitosis, which is an expression of radiation damage. Radiation can cause cell death, but slowly. Damage of connective tissue and vasculature may lead to impairment of its circulation. If the damage due to circulation is severe, secondary parenchymal cell death may occur, due to deprivation of nutrient. These damages are permanent and irreversible, and become the limiting factor of radiation treatment. The maximum dose is decided by the late tissue damage.

The localized radiation effects of few organs are explained in the following paragraphs. Most of the radiation effects on normal tissues and organ come from radiotherapy treatment experience, with partial body irradiation delivered over a period. In radiotherapy, radiation is delivered without exceeding the normal tissue tolerance. Thus, normal tissue tolerance becomes the limiting factor in radiotherapy. Hence, the observation and scoring of normal tissue reactions is an important issue in cancer hospitals. In radiotherapy, the damage to the organ is reported in terms of tumor dose (TD), which produces complications in normal tissue with 5% probability within 5 years, after the end of treatment. This is referred as normal tissue complication probability (NTCP) and its short form is NTCP $TD_{5/5}$, which is read as NTCP of 5% in 5 years. In addition, the tumor dose that produces 50% NTCP in 5 years is also used, which is denoted as NTCP $TD_{50/5}$.

SKIN

The skin is made up of epidermis, outer layer and dermis, inner layer followed by subcutaneous tissue (Fig. 7.2). The epidermis is about 100 μm thick and has 15 layers. The basal layer is the bottom layer consists of proliferating stem cells. The other layers are made up of protective dead cells, up to the surface. The dermis is below the epidermis, containing muscle fibers, blood and lymph vessels, sweat glands, hair follicles and nerves. The vascular system lies below the dermis. The epidermis is the target for radiation and causes early effects, since it is made up of dividing cells, and positioned close to the radiation. The dermis shows delayed radiation effects.

Effects of Radiation on Skin

Skin is the first organ studied for radiation effects. The effect of radiation on skin is erythema, pigmentation, dermatitis, ulceration, necrosis and malignancy. The dermatitis is further divided into radiodermatitis erythematosa, radiodermatitis bullosa and radiodermatitis escharotica. The earliest radiation effects seen in the skin are erythema and radiation dermatitis. Skin reactions are deterministic and have a threshold of 1 Gy.

Skin erythema dose: It was used as unit to deliver radiation to patients. It is defined as the dose of X-ray required to cause a certain degree of erythema within a specified time.

Early transient erythema: It occurs within 1–2 days, after an acute exposure of 2 Gy of low-LET radiation. This is clinically referred as early transient erythema. It is caused by increased capillary dilatation, permeability, and release of vasoactive amines (histamine).

Erythema stage: The erythema can reappear after 2 weeks, after repeated low exposures. The maximum response is reached at the 3rd week, when the skin appears edematous, tender and with burning sensation. This is the main erythema stage caused by inflammatory reaction and loss of epithelial cells, and its release of proteolytic enzymes. Free radical upregulates vascular damage through adhesion molecules, proinflammatory cytokines, smooth muscle cell proliferation and apoptosis.

Late erythema: It may be seen between 8 and 52 weeks after radiation exposure.

An acute radiation exposure of 3–6 Gy may cause epilation or temporary hair loss, 2–3 weeks after irradiation. The hair regrows after 2 months and gets completed within 6–12 months. An acute dose of 15 Gy or a

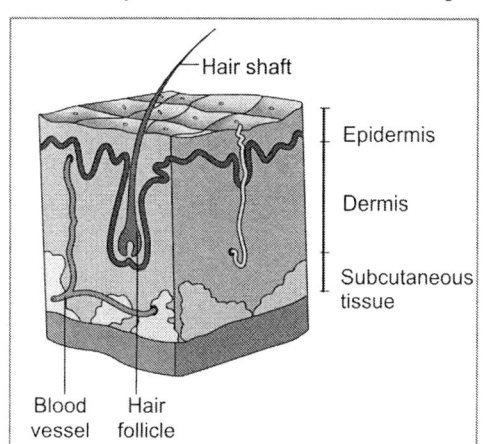

FIGURE 7.2: Layers of the skin

fractionated dose of 30 Gy may result in dry desquamation. An acute dose of 20 Gy or 40 Gy delivered over a period of 4 weeks may cause intense erythema, acute dermatitis and moist desquamation. It is clinically present as edema, dermal hypoplasia, inflammatory cell infiltration, damage of vasculature and permanent hair loss. If there is moist desquamation, it is a destruction of epidermis layer and an indicator of telangiectasia. It is a late developing vascular injury, after irradiation. The IAEA version of skin reactions are given in Table 7.1.

Table 7.1: Time of onset of clinical symptoms of skin injury related to dose (IAEA)*

Symptoms	Dose range (Gy)	Time of onset (day)
Erythema	3–10	14–21
Epilation	> 3	14–18
Dry desquamation	8–12	25–30
Moist desquamation	15–20	20–28
Blister formation	15–25	15–25
Ulceration	> 20	14–21
Necrosis	> 25	> 21

*Source: International Atomic Energy Agency (IAEA) safety reports series No. 2, 1998.

If the damage to the vasculature and germinal epithelium is less, reepithelialization occurs within 6–8 weeks, reverting the skin back to normal within 2–3 weeks. On the other hand, severe damage permits healing, and the skin may appear as atrophic, hypo- or hyper-pigmented and susceptible for physical injury. Possibility of lesions and infections are quite common in most patients, which may result in necrotic ulcer. Repeated low level radiation with a total dose of 20 Gy may result in chronic radiation dermatitis. The skin may be hypertrophic with increased risk of development of cancer. However, a chronic exposure of 6 Gy will not cause erythema.

The skin reactions are severe, if the patient is related with diabetes mellitus, systemic lupus erythematous, scleroderma and homozygosity. Apart from this, the area of radiation exposure, anatomical site and fractionation of the dose also influences the level of skin damage. The tolerance dose of skin is 60 Gy fractionated over period of 6–8 weeks.

HEMATOPOIETIC SYSTEM

The hematopoietic tissue includes bone marrow, spleen and liver. About 60% of the bone marrow is located in the pelvis and vertebrae, and the rest is in the ribs, skull, sternum, scapula and proximal sections of the femur and humerus. The liver and spleen act only during partial body exposure in an adult. Stem cells are found in the circulating system in small fractions and are radiosensitive. The stem cells go through multiplication, maturation, followed by differentiation without division, before they become mature circulating blood. The transient time from stem cell to fully functional cell differs for various circulatory blood element types. Variety of cells is present in the system that includes myelocytes, erythroblasts and megakaryocytes. They mature into polymorphonuclear leukocytes, erythrocytes and platelets. Survival curve reveals that there is no shoulder and the D_0 is less than 1 Gy. Bone marrow at the end of long bones (sternum) is more sensitive than the soft tissue.

Total Body Irradiation

A whole-body exposure of 0.3 Gy results in reduction of lymphocytes since it is the most sensitive cell in the body. A lower dose of 1 Gy may result in depression in granulocyte count, but less marked and regeneration is also less rapid. A person exposed to 1 Gy has reduced lymphocyte cell count and more susceptible to infection. Early symptoms include loss of appetite, nausea, diarrhea and vomiting. A dose of > 1 Gy may cause adverse effects to blood components. As dose increases, the effect also increases. A large dose (>1 Gy) results in alteration of blood cells, followed by lymphopenia, granulopenia, followed by thrombopenia and finally anemia.

A radiation dose of 2 Gy may result in maximal depression of blood cell count at 4–5 weeks. Megakaryocytes are more sensitive than functional platelets. Eosinophil count drops only after a large dose. Octational leukocytosis is also observed and this attributes to stress induced hyperplasia, resulting in immature cells recruited in the circulation. This may be due to radiation-induced synchrony and transient in nature. Platelets and monocytes recover slowly. Recovery of white blood cell production is more rapid than erythropoiesis. Marked anemia does not occur until 2–4 weeks after a large radiation dose owing to slow kinetics of erythrocyte depression. In case of severe damage, this turns into anaplastic anemia.

A dose of 4–6 Gy may result in temporary rise in the number of granulocytes, followed by rapid fall at the end of 1st week. The number remains constant and again falls to a minimum at 18–20 days. After 1 week of aplasia, regeneration takes place rapidly in platelets, reticulocytes and granulocytes. At higher doses, the period of aplasia is longer, susceptible to hemorrhage and infection, resulting in death. The general pattern of blood counts for modest dose of radiation is given in Figure 7.3.

Inability to carry out essential body defense mechanisms consequent upon severe damage of blood forming organs and circulating corpuscles is known to result in death usually after 2–4 weeks. The stem cell survival is critical after radiation exposure. If the number of stem cells falls below a critical level, production of functional cell stops, until partial regeneration and differentiation occurs. Hematopoietic growth factors may be administered to shorten the period of aplasia and accelerate regeneration of blood cells.

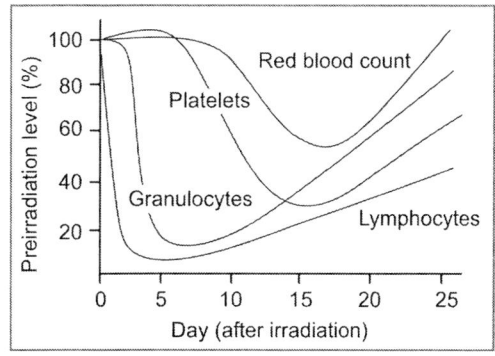

FIGURE 7.3: Blood cell depression and recovery after medium whole-body exposure

The late effects arises from the injury of blood vessels that make up the microvasculature. Damage of the capillary may affect oxygen and nutrient transport to cells, resulting in functional loss. The vessel may attain less flexibility and become stiff, resulting in decreased perfusion, which is very important in skin and brain.

Partial Body Irradiation

Lymphocytes are the most sensitive cells with a threshold dose of 0.25 Gy, hence used as biological dosimeter. The effect is transient and the count reverts to normal after a period of about 2 weeks. It is found that bilobed lymphocytes are present in circulating blood of the individuals, who are exposed to > 5 cGy. If the lymphatic count falls below 1,200/mm³ within 24–48 hours, the prognosis is serious and if it falls below 300/mm³, the individual is most certain to die. The bone marrow damage results in the rapid fall of circulating leukocytes and platelets though erythrocytes are relatively resistant. Severe damage of bone marrow is initiated at a threshold of 3 Gy in the fractionated dose.

In partial body irradiations, the stem cells in the unirradiated bone marrow start dividing within few hours, compensate hyperplasia and maintain total production of blood elements. The hematopoiesis also starts in long bones, spleen and liver, which are not normally hematopoietic in adults. As a result, the pool of stem cells in the unirradiated volume falls progressively. Doses > 30 Gy may cause permanent aplasia in the irradiated volume and hyperplasia in the unirradiated volume. Irradiation reduces the number of stem cells in the bone marrow and normally comes after a long time.

Changes have been found to exist in blood up to 3 years or more among patients locally subjected to fractionated radiotherapy. Monitoring of accidently irradiated individual's blood has been found to be a vital indicator of prognosis and also helped to initiate steps for rehabilitating them. However, the management of accident victims is a complicated one and it must be handled by a specialized team.

GASTROINTESTINAL TRACT

The gastrointestinal (GI) tract consists of oral mucosa, esophagus, stomach, small intestine and large intestine. Small intestine (jejunum) is the most radiosensitive, whereas the esophagus is the resistant one.

Oral Mucosa

The mucosa behaves like skin and its early reaction is a limitation in head and neck cancer treatment. The tissues of interest in oral cavity are tongue and tonsils. The tongue consists of muscle bundle, mucosa and taste buds. After irradiation, the muscle progress to fibrosis and fiber atrophy. Desquamation may start at day 12 of irradiation at the soft palate and spread to hypopharynx, vallecula, floor of mouth, cheek, epiglottis, etc.

The events start with hyperemia and edema after the 1st week of radiation, followed by pain and loss of interest to eat in the 2nd week. This is followed by appearance of mucositis and swelling in the 3rd week, with difficulty in swallowing. During the 4th and 5th week, the symptom progress and reaches its peak. Patient may be highly sensitive to temperature, food, touch, etc. Dysfunction of salivary gland and xerostomia are the observed clinical symptoms. However, complete recovery is possible within 2–4 weeks of post-irradiation period. The tolerance dose of $TD_{5/5}$ and $TD_{50/5}$ are 32 Gy and 46 Gy respectively.

Esophagus

The esophagus is a tube passing from mouth to the stomach and located between lungs in the chest. Its mucosa contains dividing cells; they exhibit esophagitis and increased thickness of squamous layer. The clinical symptoms are burning sensation with pain during swallowing, after 2–3 weeks of radiotherapy. The late effects include difficulty in swallowing, heartburn and sensation of lump in the throat, due to necrosis and thickening of epithelium. About 5–10 Gy is required to induce thinning of mucosa consequent upon mitotic inhibition and necrosis in the gastric wall. The tolerance dose of $TD_{5/5}$ and $TD_{50/5}$ are 55 Gy and 68 Gy respectively.

Stomach

Stomach irradiation often causes nausea and vomiting with early effects of delayed gastric emptying and epithelial denudement. Ulcer formation is a common effect on stomach, leading to bacterial invasion and rise in temperature. Late effects include ulceration in about 5 months, gastritis in about 1–12 months and dyspepsia in about 6 months to 4 years. The tolerance dose of $TD_{5/5}$ and $TD_{50/5}$ are 50 Gy and 65 Gy respectively.

Small and Large Intestine

The most radiosensitive cells in the intestine are the immature stem cells located at the base of the villi (Fig. 7.4). These stem cells develop into intermediate cells and then become mature cells, and migrate to top of the villi. During this migration, they receive

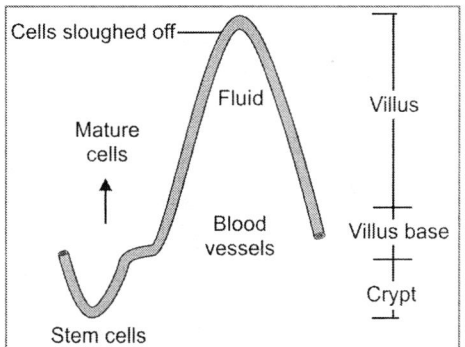

FIGURE 7.4: Intestinal villi—the most radiosensitive cells are stem cells

signal to become the functional cell of the organ. Their function includes absorption of nutrient from the gut and discharge of waste into the intestine. The matured cells at the top of the villi are soughed off at the end of their life cycle. The above whole process takes a period of about 2 weeks. Hence, the effects of radiation can be seen only after 2 weeks.

Small doses of ionizing radiation may affect enzyme secretion, whereas large doses lead to denudation of intestinal mucosa involving villi. Even small dose of 0.5 Gy can cause a drop in villi count. Damaged sites open up for invading intestinal bacteria into the blood, thus leading to sepsis. Emptying time of small intestine is reduced, thus rendering poor assimilation. Symptom includes weight loss, poor nutritional status and favoring infection to the individual. Patient may have nausea, bloody vomiting and diarrhea, after a dose of 2 Gy. However, the intestinal injury is dose rate dependent and tolerance dose for small and large intestine is different. The intestine tolerance dose of $TD_{5/5}$ and $TD_{50/5}$ are 40 Gy and 55 Gy respectively. However, the corresponding value for rectum is 60 Gy and 80 Gy respectively. Rectal complication includes diarrhea, proctitis, necrosis, stenosis and fistula.

REPRODUCTIVE ORGANS

Testes in males and ovary in females are the sites of generation of lineage of cells that go into the formation of offspring. Exposure of reproductive organs can cause tumors (somatic), impaired fertility and hereditary effects. The hereditary effects and induction of cancer are stochastic in nature, while impaired fertility is deterministic effects.

Testes

Testes contain two types of cells, namely:
- Sertoli cells
- Germinal cells.

The sertoli cells secrete hormone that controls the secretion of follicle-stimulating hormone and germinal cells define the hierarchy. Testes contain cells with differing radiosensitivity. The stem cells are more radioresistant than differentiating spermatogonia, which produces germ cells. The transient time from stem cells to spermatozoa is about 74 days in humans. The radiation effect on male reproductive system includes reduction in fertility, temporary sterility and permanent sterility. At the same time, the radiation has no effects on the leydig (libido) cells, which secrete testosterone. The threshold dose for temporary sterility is 0.1 Gy as an acute dose, with a latent period of about 3–9 weeks. For permanent sterility, it is 6 Gy for an acute radiation, whereas it is < 6 Gy for fractionated radiation, with a latent period of about 3 weeks.

There will be a window period for fertility before the onset of sterility. The recovery begins at 1 year after radiation exposure of 1 Gy; however, the sterility duration is dose dependent. This is due to the sequence of germ cells, i.e. spermatogonia, spermatids, spermatocytes and mature spermatozoa, which are available as pool. The mature spermatozoa are the most resistant to radiation. Only the sensitive precursor cells are selectively killed. Most of the cell killing is due to intermediate stages of spermatogenesis, which are radio sensitive that results in sterility. Permanent sterility may appear for fractionated radiation doses of 20–50 mGy per week totaling to 2.5–3 Gy. The causes for the reduced fertility is the decreased sperm count (oligospermia) and motility (asthenozoospermia), which can occur at about 6 weeks after a dose of 150 mGy.

Fractionated radiation or continuous LDR radiation is more effective than single acute exposure. Mostly, the stem cells are in the radio resistant 'S' phase of the cell cycle. If the radiation is fractionated, the stem cells moves to the radiosensitive phases and are effectively killed. This may be the cause for the longer period of azoospermia in patients underwent fractionated low-dose pelvic irradiation and testicular dysfunction, found in occupational radiation workers. The tolerance dose of $TD_{5/5}$ and $TD_{50/5}$ are 1 Gy and 2 Gy respectively.

Ovary

The effect of radiation on ovary is different from testes. The oocytes are present from birth and no longer divide further. Their number decreases with age and reaches zero at the time of menopause. The oocytes are sensitive to radiation and undergo interphase death. The ova within ovarian follicles are sensitive to radiation. The intermediate follicles are the most radiosensitive, followed by mature follicles and small follicles. A radiation dose of 1.5 Gy may cause temporary sterility with reduced fertility.

Permanent sterility may occur at 3 Gy for acute doses with a latent period of less than 1 week. It is 6 Gy for fractionated radiation dose. The permanent sterility is age dependent and people in the age group of 35–40 are more susceptible than young women. In adult females, the process of ovum formation is much simpler owing to the presence of relatively differentiated oocytes. Hence, there is no apparent latent period for sterility induction. Because of the above reason, the ovary is more resistant than testes.

The reproductive organ of the fetus is formed in the first month of pregnancy, and hence, irradiation of the embryo results in reduction of fertility or in sterility throughout the life time. The radiosensitivity of the gonads of the fetus varies with state of development of the embryo and the early fetal stage is most sensitive. The tolerance dose of $TD_{5/5}$ and $TD_{50/5}$ are 2–3 Gy and 6–12 Gy respectively.

EYE LENS

The eye lens has no blood supply and consists of population of radiosensitive cells. Cell division continues throughout the life, but there is no mechanism for removal of dead cells. Radiation can damage these cells and cause cataract or opacity of the lens (abnormal lens fibers). This is the result of abnormal differentiation of damaged epithelial cells, which accumulates at a point and cause vision impairment. Usually, these cells accumulate in the anterior subcapsular region and migrate posteriorly. The posterior subcapsular cataract causes glare or halos around light at night, resulting in vision impairment.

This cataract was thought of a deterministic effect and the threshold for acute and chronic radiation exposure are 2 Gy and 5–8 Gy respectively for X-rays with a latent period of about 2–3 years. Based on the concept, the ICRP (report 60, 1990) fixed the annual dose limit as 150 mSv per year. However, recent studies have concluded that the threshold dose values for both acute and fractionated radiation is of the order of 0.5 Gy, with a latent period of more than 20 years. There is also suggestion that the dose response of eye lens is better explained with linear non-threshold stochastic model. On the basis of above, the ICRP (2011) has revised the annual dose limit as 20 mSv per year instead of 150 mSv per year for radiation workers. It is the average dose over defined periods of 5 years, with no single year exceeding 50 mSv.

The physicists worked with cyclotron experienced cataract due to densely ionizing radiation of neutron and proton. High-LET radiation is more effective for cataract induction, as found in neutron. The onset of cataract is found to be dose and dose rate dependent, and the latent period is inversely related to dose. The tolerance dose of $TD_{5/5}$ and $TD_{50/5}$ are 50 Gy and 65 Gy respectively, and higher dose may lead to blindness.

OTHER TISSUES

Lungs

Lungs are the most radiosensitive organ in the thorax, and hence, special care is taken to exclude them from the radiation field during radiotherapy. Radiation can lower the amount of lung surfactant, by which lung expands. This may create dry cough and shortness of breath. Radiation pneumonitis is an inflammatory response, can be observed in 1-6 months after radiotherapy. The symptoms are fever, cough and shortness of breath. Pulmonary fibrosis, the formation of scar tissue in the lungs can occur with symptoms of shortness of breath and decreased ability to exercise. This may be a permanent feature in patients. The tolerance dose of $TD_{5/5}$ and $TD_{50/5}$ are 17.5 Gy and 24.5 Gy respectively.

Kidney

Kidney is radiosensitive and late responding critical organs. A radiation of dose 30 Gy in 15 fractions may result nephropathy with arterial hypertension and anemia. Other clinical complication includes nephritis that occurs after a month and kidney failure. A single dose of 6 Gy results in an early damage and a dose 10 Gy results in protein spilling into urine. The tolerance dose of $TD_{5/5}$ and $TD_{50/5}$ are 23 Gy and 28 Gy respectively.

Liver

Liver has enormous reserve capacity and it is able to maintain apparent normal function, even part of it is being damaged. It becomes important only if the whole liver is irradiated, e.g. whole-body irradiation. This does not get exposed severely in radiotherapy, before considerable damage is done to other sensitive tissues, because of its location. The life span of hepatocyte is about 1 year and the cell renewal rate is very slow. Hence, liver is considered as radioresistant tissue. It tolerates even a large dose, however, hepatic function may deteriorate. The tolerance dose of $TD_{5/5}$ and $TD_{50/5}$ are 30 Gy and 40 Gy respectively.

Bladder

Bladder tissue contains of diploid, transitional and polyploidy cells. The cell renewal rate is slow in the epithelium and cell's life span is long (months); hence, there is delayed proliferation after irradiation. Increased frequency of urination and irritation by urine has been reported. The late effects include fibrosis and reduction in bladder capacity. The tolerance dose of $TD_{5/5}$ and $TD_{50/5}$ are 65 Gy and 80 Gy respectively.

Brain

Brain is morphologically resistant to ionizing radiation, but functional changes are known to occur often at low doses. Since brain cells do not reproduce, they are not significantly damaged, until their blood supply is disturbed. The important injuries are necrosis, infarctions and late syndromes. Infarction is a condition for lack of oxygen, caused by obstruction of blood vessels. Histopathological changes may occur in the white matter after in 1 year and later spread to grey matter and vascular regions. Brain necrosis has been reported in irradiated humans, involving white matter undergoing demyelination. The tolerance dose of $TD_{5/5}$ and $TD_{50/5}$ are 45 Gy and 60 Gy respectively.

Spinal Cord

Spinal cord and brain forms part of central nervous system, which is less sensitive to radiation. Radiation-induced effects on spinal cord are similar to that of brain. The complications are myelitis, necrosis and paralysis, which can increase with the length of the cord irradiated. Early demyelination may occur at a dose of 35 Gy. The late effects include two syndromes, namely demyelination and necrosis after 6–18 months and vasculopathy in 1–4 years. The tolerance dose of $TD_{5/5}$ and $TD_{50/5}$ are 47.7 Gy and 68.3 Gy respectively, for 20 cm cord length. In the case of late damage $TD_{5/5}$ is about 50 Gy for a 10 cm length spinal cord. The tolerance dose has small dependence on overall treatment time. However, dose per fraction has impact on time, and hence, much caution is required in hyperfractionation types of treatment.

Heart

Radiation exposure may damage the heart muscle, valves or coronary arteries. The heart muscle damage often called cardiomyopathy, involves stiffness of ventricle, which do not respond to signals related blood pumping. This may lead to reduced blood pumping and reduced oxygen to the heart, resulting in congestive heart failure. After radiation exposure, the heart valves become stiff and do not seal properly, resulting in leakage of blood back into the heart. This will reduce the output blood to other parts of the body.

Radiation-induced coronary artery effects arise from damage of small blood vessels. Radiation can roughen the inside parts of the small vessels and permit formation of fatty deposits or plaque. Additionally, calcium deposit may harden the plaque, resulting in state of atherosclerosis. The plaque can be clogged in one of the heart vessels, ending up with coronary artery disease. In this stage, the heart muscle does not get enough oxygen and nutrition, and becomes week and may die shortly. If the heart beat faster with this condition, angina may occur. If the heart vessel is completely blocked, section of the heart muscle may die and the patient may have minor heart attack, sometimes life-threatening. The tolerance dose of $TD_{5/5}$ and $TD_{50/5}$ are 40 Gy and 50 Gy respectively.

ACUTE RADIATION SYNDROME

Whole-body exposure of large dose of ionizing radiation in short burst, over a few minutes to an hour, causes severe damage to all the organs of the body. The composite effect resulting from the damage of different system in an organism is referred to as acute radiation syndrome or sickness. This is common in reactor accidents and nuclear war or atomic bomb blast. However, this is nothing to do with medical use of radiation, which is basically partial body irradiation involving low dose and fractionation. Acute radiation syndrome is basically deterministic effects that arise from the loss of tissue or organ function. Hence, it always occurs with threshold dose. These syndromes develop not only due to the effects on specific organ but also due to the interaction of the effects produced in different organs. Following a radiation dose, variety of syndrome occurs in most mammals. The time of appearance is characteristic of the specific syndrome. The incidence of radiation sickness is also found to be dependent upon the part of the body and extent of irradiation. For example, exposure over the whole trunk and particularly over the upper abdomen causes more radiation sickness than dose exposure of comparable tissue volumes in the extremities. The resources of acute radiation effects are obtained from the animal experiments, Hiroshima and Nagasaki atomic bomb survivors (1945) and Chernobyl Nuclear Power Plant Incident (1986). As on date, about 400 humans have died due to acute radiation syndromes.

In addition, there are two more incidents, one involving abandoned teletherapy machine, cesium-137 source of 50.9 TBq activity, in Goiania, Brazil (1987). While removing the source, it gets ruptured and people are exposed to large dose of external radiation as well as internal contamination. In this, 4 people died and 28 people developed local radiation injury. The second event involves industrial sterilization facility, in San Salvador and El Salvador. The cobalt-60 source stuck in the opening position and three workers were exposed to high level of radiation. However, medical management has saved their life with amputation of legs in two workers. One worker has died after 6.5 months due to lung damage. Based on the above data, the condition for acute radiation syndrome can be summarized as follows:

- Large dose of ionizing radiation
- External irradiation
- Highly penetrating radiation such as X- and gamma (γ)-rays
- Whole-body exposure
- Short delivery of dose.

There was increasing interest to know more about the dependence of latency to death on whole-body radiation dose. Hence, series of animal experiments were conducted with mice and dogs. For a whole-body dose range of 5-12 Gy, the survival decreases gradually in such experiments. Increasing the dose beyond 30 Gy did not decreases the latency to death. However, further increase in dose > 30 Gy caused death within few days or within few hours. The above such pathogenic happenings on humans are divided into three types of syndromes, namely (Fig. 7.5):

- Hematopoietic syndrome (< 12 Gy)
- Gastrointestinal syndrome (12–30 Gy)
- Neurovascular syndrome (> 30 Gy).

The acute radiation syndrome takes place in steps over a period. This period may vary from hours to days or weeks. This is due the variation of sensitivity of various organs and tissues in the body. The nature of the syndrome

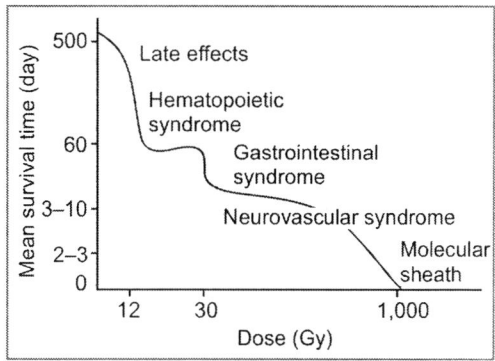

FIGURE 7.5: Radiation syndromes are related with whole-body radiation exposure

depends upon the magnitude of the acute dose, involving whole-body exposure of X- or γ-rays or neutron.

Though the concept of syndrome is evolved from animal experiments, the present understanding is based on careful clinical evaluation of human accident victims from:

- Oak Ridge accident (1958)
- Chernobyl disaster (1986)
- Tokai-Mura accident (1999).

Mean Lethal Dose

After the fatality of human population in Hiroshima and Nagasaki, several animal experiments were initiated to find the relation between lethality and radiation dose. Experiments from mice to goats and dogs revealed that the lethality increases with increase of dose in the sigmoid dose-response curve. The radiosensitivity is usually defined by the steepest part of the dose-response curve indicating lethality in 50% subjects in a specific period. The period is usually 30 days.

Thus, the radiation death end point is scored on the basis of 50% lethal dose (LD_{50}), called mean lethal dose. It is defined as the dose required to causes fatality in 50% of the population within 30 days or simply known as $LD_{50/30}$. The time period in which the death is assessed is referred by a subscript, denoting the number of days that has elapsed. Usually, various syndromes are related to various time periods. The mean lethal dose vary with species and is not absolute within the species. It depends on the body weight and general health of the individual exposed. There is a decreasing trend $LD_{50/30}$ with increasing body mass. It varies from 6.2 Gy for mice to 2.5 Gy for larger animals. The mean lethal doses of various species are shown in Table 7.2.

Table 7.2: $LD^*_{50/30}$ values for various species irradiated to whole body acutely by X- or γ-rays.

Species	$LD_{50/30}$ (Gy)
Pig	2.5
Dog	2.75
Guinea pig	3.0
Monkey	4.25
Human	3.5–4.5
Mouse	6.2
Gold fish	7.0
Hamster	7.0
Rat	7.1
Rabbit	7.25
Gerbil	10.5
Turtle	15.0
Newt	30.0

*LD, lethal dose

In other words, $LD_{50/30}$ is a guide to assess the prognosis of the exposed individual or animal. There is another way of expressing lethal dose called $LD_{50/60}$; it is defined as the dose to kill 50% of the exposed population within 60 days. Based on the human population deaths from the Hiroshima and Nagasaki, the $LD_{50/60}$ is found to be 3–4.5 Gy for humans. The $LD_{50/30}$ for human is about 4 Gy without medical management after the exposure, however the clinical symptoms vary with individuals. The lethality in human more depends on other factors such as:
- Co-morbidity
- Quality of medical care.

Hence, three ranges of doses are defined, instead of $LD_{50/30}$, by the medical research council, UK. They are:
1. Survival very likely ≥ 2 Gy.
2. Survival possible with adequate medical care < 2–8 Gy.
3. Survival unlikely with adequate medical care ≥ 8 Gy.

It is well understood that most of the radiation effects are due to radicals and oxygen effect in species rich of water. In flies and insects, the water content is less and they are radio resistant. Hence, radiation dose of several kGy is required in sterilization and food preservation.

Syndrome Events

In general, the sequence of events of acute radiation syndrome after the whole-body radiation is described as (Fig. 7.6):
- Prodromal period
- Latent period
- Manifestation of illness
- Death or recovery.

The prodromal radiation symptoms occur soon within minutes and hours after high-intensity radiation. A whole-body exposure of 0.5–1 Gy may cause prodromal symptoms that include anorexia, nausea, vomiting, diarrhea, fever, lymphopenia and granulocytosis (Table 7.3). The symptom of prodromal period is generally divided into:

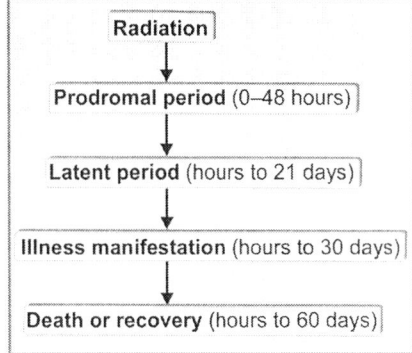

FIGURE 7.6: Sequence of events that is taking place after an acute radiation exposure

Table 7.3: Clinical findings of acute radiation syndrome during the prodromal period (IAEA)*

Symptoms (onset)	Mild (1–2 Gy)	Moderate (2–4 Gy)	Severe (4–6 Gy)	Very severe (6–8 Gy)	Lethal (> 8 Gy)
Vomiting	2 h	1–2 h	< 1 h	< 30 min	< 10 min
Diarrhea	Nil	Nil	Mild	Heavy	Heavy
Headache	Slight	Mild	Moderate	Severe	Severe
Consciousness	Normal	Normal	Normal	May be altered	Unconscious
Temperature	Normal	Increased 1–3 h	Fever 1–2 h	High fever < 1 h	High fever < 1 h
Medical treatment	Outpatient	Admission and observation	Admission + Specialized treatment	Admission + Specialized treatment	Palliative treatment

Source: International Atomic Energy Agency (IAEA) safety report series No. 2, 1998

- Gastrointestinal
- Neuromuscular.

The prodromal symptoms vary with time of onset, maximum severity, duration and it depends on the dose. The laboratory tests such as lymphocyte count and chromosome aberration assay are the best techniques to establish prodromal symptom of an individual.

Prodromal is followed by a latent period, during which the prodromal symptoms subside. The duration of the latent period is shorter for higher doses and longer for low doses. Thus, the latent period duration is inversely proportional to the dose. Absence of latent period reveals that the dose is very high. For a modest dose of 1 Gy, the latent period is about 4 weeks. This is followed by illness manifestation, which lasts for 2–4 weeks or even longer. In this stage, the hematopoietic system gets damaged and results in immunoincompetence. Hence, it is important to offer effective treatment for about 6–8 weeks after exposure. If the victim survive this state, recovery is possible. However, there is a higher risk for cancer and genetic abnormalities of the progeny.

Systemic Involvement

Hematopoietic Syndrome

A whole-body dose of about 0.5–10 Gy of acute radiation may cause hematopoietic syndrome. A whole-body dose of 0.7 Gy result in, destruction of blood producing bone marrow and manifest fatigue with symptoms of anemia, anorexia and vomiting. If the person is exposed to 1 Gy, the lymphocytes count reduces and the body is susceptible for infection. The early symptoms are loss of appetite, diarrhea and vomiting. If the dose is < 1–2 Gy, all components of the blood get affected. The white blood cells (WBCs) count comes down, and damage to the skin and intestine make the person more prone to infection. This will happen within 2 weeks after the exposure and the individual can be recovered from the symptoms, if medical treatment is offered.

If the dose is > 2.5 Gy, the supply of mature red blood cells (RBCs), WBCs and platelets are severely damaged. The blood cells supply will not be maintained properly, due to the damage of hematopoietic stem cells. Death may occur within few months due to infection, anemia and hemorrhage. If the dose is > 8 Gy, the individual usually end up with death. However, medical treatment can be attempted by offering colony stimulating factors or bone marrow transplantation or stem cell therapy. If the dose is > 12 Gy, the patient may not survive even with medical treatment, due to irreversible damage of GI tract and the vasculature.

The prodromal symptoms occur within few hours after exposure that includes nausea, vomiting, headache and diarrhea. Early prodromal symptoms and severe diarrhea occur within 2 days and may cause death. The prodromal symptoms and latent period exists for weeks. During the latent period, nausea and vomiting may subside, but patient may feel fatigue. The stem cell number reduces and it is unable to maintain the normal hematologic profile. The lymphocyte counts decreases within hours after exposure and it can be used as a biological dosimeter. The threshold value for a measurable depression in the blood lymphocyte count is about 0.25 Gy. Lymphocyte count less than 1,000/mm^3 in the first 48 hours will indicate overexposure.

The clinical symptoms appear as depletion of bone marrow, hemorrhage due to platelet loss, severe neutropenia and prone to infections, etc. The systemic effects include immunologic compromise, sepsis, hemorrhage, anemia and impaired wound healing. The cause of hematopoietic syndrome is found to be sterilization of acute precursor cells by the radiation. This decreases the supply of mature RBCs, WBCs and platelets. Thus, it reaches the minimum value over a few weeks. The supply of new cells from the precursor is inadequate and worsens the condition.

In human population, the radiation-induced hematopoietic effects are not uniform and are heterogeneous in nature. The young and old people appear to be more radiosensitive than adults. The females are more radioresistant compared to males. The signs of hematopoietic damage appear slowly in humans compared with animals. The death starts at 30 days and continue up to 60 days. Hence, the LD_{50} for humans is expressed as $LD_{50/60}$ and its range is 3.5–4.5 Gy. This may be extended up to 5–6 Gy with supportive treatments such as transfusions and antibiotics. Use of hematological growth factors further may increase its value to 6–8 Gy. In the case of mice, it is written as $LD_{50/30}$, as death occurs between 10–30 days. During the Chernobyl accident, about 203 individual had suffered from hematopoietic syndrome, after an exposure of about 1 Gy. Out of which 13 individuals have died and the rest of them are recovered from the effects.

Gastrointestinal Syndrome

A whole-body dose of 12 Gy of gamma radiation may cause GI syndrome. In GI syndrome, death may occur in 3–10 days, if no medical treatment is given. The damage of GI tract occurs even at low doses of 2–10 Gy, which is known for the occurrence of hematopoietic syndrome. The various prodromal symptoms are nausea, vomiting, anorexia, prolonged diarrhea and cramps within hours. This is due to injury of epithelial lining of the GI tract, which prevents normal food intake and fluid intake, resulting electrolyte imbalance. Persistent diarrhea may progress from loose to watery and then bloody. The prodromal symptoms repeat along with intestinal dysfunction, due to damage of intestinal mucosa. Death may occur within a week due to infection, dehydration and electrolyte imbalance. The latent period is shorted of the order of 5–7 days.

The causes for GI syndrome are depopulation of epithelial lining of GI tract by the radiation. The normal intestinal lining is made up of immature stem cells, differentiating cells and mature functional cells. The immature stem cells are present at the base of the villi in the intestine. They maintain the supply of stem cell for further development of differentiated cells and matured functional cells. The matured stem cells perform the functions of absorption of nutrients from the gut and discharge of waste products into the intestine. At the end of their life cycle, the matured functional cells at the top of the villi are soughed off. The whole process require at least 14 days.

The most radiosensitive cells are the immature stem cells at the base of villi. Radiation exposure depopulate the immature stem cells in villi and the villi begin to shrink over a period of few days and finally the intestine is denuded of cells. The denuding of intestinal cell may cause pathophysiological sequelae. The intestine no longer maintains the barrier between body waste inside the intestine and bloodstream. This breakdown of the mucosal barrier allows the entry of luminal contents including antigens, bacteria and digestive enzymes into intestinal wall, resulting in radiation-induced intestinal mucositis. This reduces the capacity to regulate absorption of electrolytes and nutrients. A portal is created for intestinal flora to enter the systemic circulation. The changes in the bone marrow are also added up with the GI tract changes. Hence, there is a severe reduction of white cells circulation, while bacteria are entering the blood stream from the GI tract. The rate of cell loss may be slow or fast that depends on the dose.

Clinical symptoms of intestinal pathology: This includes mucosal ulceration and hemorrhage, disruption of normal absorption and secretion, alteration of enteric flora, depletion of gut lymphoid tissue and disturbance of gut motility. The systemic effects in the individual includes malnutrition due to malabsorption, vomiting and abdominal distension from paralytic ileus, anemia from GI bleeding, and dehydration and acute renal failure due to fluid and electrolyte imbalance.

It is not affecting the differentiated and functioning cells, since it requires large dose of radiation. Since the functional cells are not affected, the radiation effects are not felt immediately. The time delay between the time of irradiation and the onset of syndrome is decided by the life span of mature function cells. The time schedules of events vary with the type of organism. It is common that the individual suffering from GI syndrome has been already with hematopoietic syndrome. During the Chernobyl, several fire fighters have died within 10 days after the accident, due to GI syndrome.

Neurovascular Syndrome

The neurovascular syndrome is also known as central nervous syndrome. A whole-body dose of 30 Gy of gamma radiation may cause death of an individual within 3 days. This is due to the result of cardiovascular shock with loss of serum and electrolytes into extravascular tissues. The occurrence of edema, elevated intracranial pressure and cerebral anoxia cause death before the damage of other organs. The death mostly occurs due to the destruction of the whole body circulatory system and increased pressure on the brain, due to fluid leakage.

The prodromal symptoms start with burning sensation of skin within minutes, followed by nausea, vomiting, confusion, ataxia and disorientation within 1 hour. The latent period is 4-6 hours during which some improvement is there. The prodromal symptoms return again severely, along with respiratory distress and cross neurologic changes. This involves tremors and convulsions resulting to coma and death.

In the dose level of 50 Gy, neurovascular syndrome always associates with other GI and hematopoietic syndromes. Since the neurovascular syndrome occurs so quickly, the symptoms of other syndromes and failure of other organs may not be observed.

The level of symptom varies with organism and the magnitude of the dose. The mechanism of neurovascular syndrome is not fully understood in humans. However, it is believed that occurrence of hypotension due to release of histamine and greater damage of microcirculation.

At very high doses of the order of 1,000 Gy, death may occur even during irradiation. This is called molecular death, because such massive doses probably will cause inactivation of many vital substances, which are needed for the basic metabolic process of the cells and tissues.

Management of Radiation Accidents

The medical management of radiation accident employs METROPOL system. The METROPOL stands for 'medical treatment protocols', as suggested by the European Consortium of Expert Scientists (2001). It defines the response criteria for four organ system with signs and symptoms. The four organ systems are (Table 7.4):

- Neurovascular system (N)
- Hematopoietic system (H)
- Cutaneous system (C)
- Gastrointestinal system (G).

Each organ system severity is divided into four grades, namely mild, moderate, severe and fatal.

RADIATION EFFECTS ON EMBRYO AND FETUS

Radiation exposure to the pregnant women simulates the condition of in utero exposure that will affect the embryo and fetus. Human data is available from the atomic bomb survivors of Hiroshima and Nagasaki, and mothers received pelvic radiotherapy before realizing that they were pregnant. About > 1,500 children born out of mothers exposed to a dose of 0.1–1 Gy in intrauterine have been studied for late effects in Japan. Only microcephaly and severe mental retardation is found in 18 children. This includes 15 mothers exposed during 8–15 weeks of pregnancy and 3 were exposed at later period of pregnancy. In general, radiation can produce:

Table 7.4: Organ systems and remedial measures (IAEA)*

Grade	Description	Remedial measures
Neurovascular system		
N1	Late onset of mild prodromal and mild fatigue symptoms, persist for several weeks	Antiemetic treatment on outpatient basis
N2	Vomiting during prodromal phase and moderate fatigue, lasting several weeks	Antiemetic treatment and regular clinical monitoring
N3	Severe nausea and vomiting within 1 hour, lasting for 2 days; symptoms may recur—leads to electrolyte imbalance; headaches, severe fatigue syndrome, hypotension and fever	Require admission, administration of intravenous glucocorticoids, electrolyte and fluid replacement and analgesics
N4	Severe nausea, vomiting, headaches, fever, erythema and drowsiness within 1 hour, recovery unlikely	Fluid and electrolyte replacement, analgesic medication, intravenous (IV) glucocorticoids and mannitol infusion to reduce intracranial pressure

Contd...

Contd...

Grade	Description	Remedial measures
Hematopoietic system		
H1	Cell count just below the lower end of the normal count range	No treatment is required
H2	Lymphocyte count on day 2 is 500–1,500/µL by transient granulocytosis; decrease to the lower end of the normal level until day 10, followed by abortive rise Granulocytopenia occurs between 12 and 20 days, < 1,000/µL result in infection Regeneration starts after 30 days; platelets decreases gradually < 50,000/µL over 3 weeks resulting in hemorrhage	Patients prone to infection and hemorrhage require treatment with antibiotic or platelet transfusion
H3	Lymphocyte count on day 1–3 is 250–500/µL Decrease in platelet starts after day 5 and become 500/µL around 10–15 days, regeneration starts after 30 days; platelet starts decreasing from 3rd week	Similar treatment of H2; however, simulation of hematopoiesis with growth factors may be considered; platelet transfusion and granulocyte colony-stimulating factors are recommended A single IV injection of 5 µg/kg thrombopoietin may give good response over 10–14 days
H4	Lymphocytes < 250/µL, granulocytes < 500µL after 1st week of exposure and platelet become 0 after 10 days	Stem cell transplantation
Cutaneous system		
C1	Early transient erythema of the skin, subside within 36 hours; a second erythema appears after 5 days; skin appear dry after 3–4 weeks, due to loss of sebaceous glands and may associate with mild pain and itching	Anti-inflammatory lotion or powder and symptomatic treatment
C2	Erythema progress to edema and blistering after 5–10 days, within 10% of the body surface, followed by transient loss of hair after 14 days	Treatment with topical glucocorticoids, linoleic acid creams and systemic antihistamines
C3	Same signs of C3, but extend to 10–40% of body surface; risk of deeper ulceration with concomitant hematopoietic syndrome	Systemic treatment with glucocorticoids and analgesics
C4	Same signs and symptoms of C3, but extend to > 40% of the body surface; it involves deeper underlying tissues of the skin and subcutaneous tissue	Require intensive care treatment to tackle pain, infection and necrosis; survival patients if any, may have long-term skin damage

Contd...

Contd...

Grade	Description	Remedial measures
Gastrointestinal system		
G1	Episodes of few altered stool consistency and frequency with associated abdominal pain	Require no treatment
G2	Change in frequency, consistency and blood in stool—both with abdominal cramps	Spontaneous recovery is possible; however, Loperamide treatment for diarrhea
G3	Higher frequency of the above episodes per day over several weeks	Spontaneous recovery is possible, but treatment for electrolyte imbalance is required; antibiotics, anti-inflammatory drugs and analgesics may be opted based on clinical symptoms
G4	Rapid onset of diarrhea may be explosive Frequent diarrhea leads to fluid and electrolyte imbalance with painful abdominal cramps; simultaneous granulocytopenia may occur leading to septicemia	Symptomatic treatment, fluid and electrolyte replacement, and antibiotics and analgesics

Source: International Atomic Energy Agency (IAEA), 2000

- Lethal effects
- Malformations
- Growth delay and retardation.

The embryo is a developing organism and undergoes rapid cell proliferation, migration and differentiation. Hence, it is the most sensitive to ionizing radiation as stated by the Bergonié-Tribondeau law. The radiation response depends on:
- Radiation quality
- Total dose
- Dose rate
- Stage of development at exposure.

The above factors altogether exhibit effects such as prenatal death, congenital anomalies, growth impairment, intelligence reduction and genetic alterations. The gestation period is divided into (Fig. 7.7):
- Preimplantation
- Organogenesis
- Fetal period.

Preimplantation refers to the time from fertilization to attachment of zygote to the uterine wall, organogenesis is the period of organ development and fetus refers to the growth of structures.

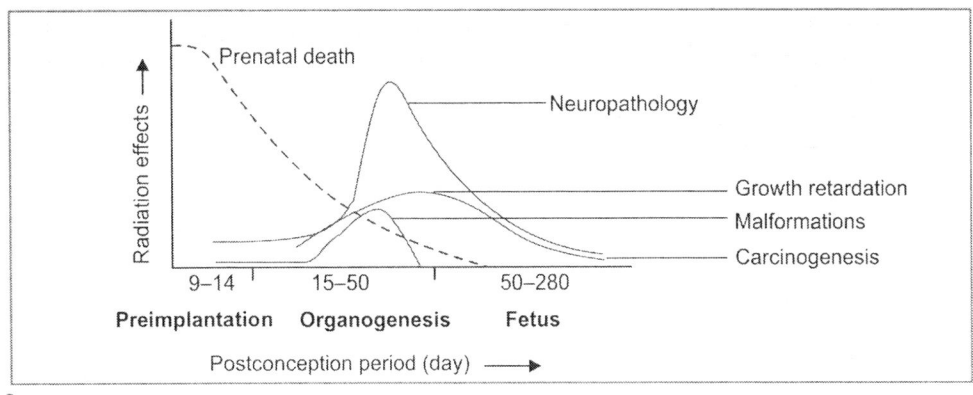

FIGURE 7.7: Health effects of radiation exposure in utero at various stages of gestation

Preimplantation

In preimplantation stage, the sperm and egg get united and forms the zygote, which is embedded in the uterine wall. The period of this stage is 9–14 days after conception. In this period, two pro nuclei fuse together, cleave and create the morula and blastula. The newly formed conceptus is sensitive to radiation damage and susceptible for lethal effects. Preimplantation is the most sensitive stage to the lethal effects of radiation. However, for a radiation dose of < 100 mGy, such lethal effects are low or negligible in humans (ICRP, 103). If the embryo survives, it grows normally. The radiation damage is repaired and there is no additional radiation-induced congenital anomaly. The causes for the embryo resistance to abnormalities are repair capacity, lack of cellular differentiation, hypoxic state, etc. Thus, the irradiated embryo survives and grows, and this is called all-or-nothing effect of radiation. This is because that the damage is likely in the form of failure to implant or undetected death in the conceptus.

It is also stated that the cells are undifferentiated and lack determination during the first few divisions. The radiation exposure may kill only some cells and the remaining could continue their embryonic development. If too many cells are killed, the embryo dies and is reabsorbed. If any misrepair damage is there, it will be exhibited later. When cells are specialized and determined, even a loss of few cells may lead to growth retardation or prenatal death. The most sensitive time in humans is:
- 12 hours after conception, during which the two pronuclei fuse to one-cell stage
- The first two cell division period that occurs at 30–60 hours after conception.

Any chromosomal aberrations at the one cell stage may result in loss of chromosome in further cell division, which become uniform throughout the embryo. Chromosomal loss is said to have lethal effects. If there is a loss of sex chromosome in female, it may result in Turner's syndrome. The women will usually be unable to realize that she is in preimplantation stage. A dose of 50–100 mGy of ionizing radiation is found to increase spontaneous abortion in animals. To induce preimplantation death after preimplantation period, a dose of 250 mGy is required. A dose of 10 mGy to the conceptus may increase the spontaneous abortion rate by 1%, whereas the natural abortion frequency in human is 30–50%.

Organogenesis

The period of organogenesis is said to be 15–50 days after conception. In this period, the differentiated cells form certain organs on a specific gestation day. That is neoplasms occur on 18th day, forebrain and eyes begin to form on 20th day, etc. Hence, each organ is not having equal risk, during this period. Radiation exposure may cause malformation in specific organs, if it is exposed during its peak differentiation period. There is also chance of damage in adjacent organs, which may have negative effect on organ development.

Usually anomalies have one or two critical periods, e.g. cataract has 2 critical periods, as reported in mice study. Animal experiments conclude that malformation has a threshold dose of 100 mGy. The most sensitive organ towards low-LET radiation is the central nervous system (CNS) below the dose of 250 mGy. Embryo exposed to radiation in early organogenesis exhibit intrauterine growth retardation due to cell depletion. There is a relation between growth retardation and teratogenic effects. Atom bomb survivors, who has received neutron and gamma dose of 100–200 mGy, have exhibited increase of microcephaly. The gestational period in humans is shorter than animals; hence, the teratogenic effects are lesser in humans. However, the CNS development takes place over a longer gestational interval than animals. Hence, CNS becomes the suited target for low-dose radiation in humans. A dose of

< 250 mGy may cause CNS malformations in humans. Exposure during the early organogenesis period may result in intrauterine death and growth retardation.

Fetal Period

The stage of fetus formation is between 50 and 280 days. The congenital anomalies and prenatal death are unlikely during this period. The primary target for radiation is the nervous system and sense organs. It may affect the growth and development of the organs. Evidence suggests that the exposed fetus may have smaller head, smaller brain size and display cognitive deficits associated with delayed brain development, behavioral alterations or reduced intelligence in later part of life. The fetus is more sensitive to mental retardation during the 8–16 weeks of pregnancy, which is the time of rapid development of brain.

However, there is no such evidence available in humans for a dose of < 100 mGy. Much higher doses are required to cause lethality during this period. The reduction in IQ is estimated as 30 IQ points per 1 Gy dose during this period. Another radiation effect is the childhood cancer due to in utero exposure. The 'Oxford Survey of Childhood Cancer' reveals that there is an association between the induction leukemia and in utero X-ray exposure. Doll and Wakeford (1997) observed that the obstetric X-ray examination in the third trimester increases the childhood cancer risk by 40%. The risk is increased by a dose of 10 mGy and the absolute risk is about 6% per Gy. The rate of cancer death in children is double the rate of cancer death in adults. In the case of medical exposures, the radiation dose to fetus is about 25% of the entrance skin dose.

Termination of Pregnancy

Doses lower than 100 mGy generally carry negligible risk during the preimplantation and organogenesis period. Even the risk is very low in the fetal period for a dose of 200 mGy. If the fetal dose is > 500 mGy at 7–13 weeks, there is a risk of intrauterine growth retardation (IUGR) and CNS damage. According to ICRP-84, termination of pregnancy at fetal doses of less than 100 mGy is not justified based upon radiation risk. At fetal doses between 100 and 500 mGy, at 7–13 weeks, the decision should be based upon the individual circumstances. Of course, this is a guideline value and other parameters must be taken into account with parental decision in consultation with the referring physician. This includes pregnancy hazard to mother, the expectancy of parents, their mental outlook, and ethnic and religious family background. It also requires the provision of counseling for the patient and her partner.

Diagnostic X-rays

The fetal dose from medical diagnostic procedures rarely exceeds 50 mGy. This will not put the fetus any significant increase in risk for congenital malformation or growth retardation. At fetal doses exceeding 500 mGy, there can be a significant fetal damage, the magnitude and type of which is a function of dose and stage of pregnancy. Many believe that this dose can cause sterility in the exposed individual, but really it is not so. The gonads are radiosensitive organs in the human body. The threshold radiation dose for permanent sterility in men is 3,500–6,000 mGy and for women 2,500–6,000 mGy. It is estimated that the excess risk of childhood cancer from in utero irradiation is approximately 6% per Gy.

Nuclear Medicine

Radiopharmaceuticals administered to pregnant women are associated with radiation risk to the fetus. There are two types of radiopharmaceuticals:
- Those that cross placenta
- Those that remain on the maternal side.

For example, radioiodine can cross the placenta and concentrate on the fetal thyroid after 13th week of gestation. It may be 55–75% between 14 and 22 weeks of gestation. This may result in hyperthyroidism or ablation. Estimates show that the dose to the fetal thyroid may range 230–580 mGy/MBq for gestational period of 3–5 months. Total body fetal dose estimate ranges from 0.072 to 0.27 mGy/MBq during pregnancy.

Most female patients are advised not to become pregnant for at least 6 months after radiotherapy with radioiodine. This is not based upon potential heritable radiation effects, but rather upon the need to be sure that:
- Hyperthyroidism or cancer is controlled
- Another treatment with radioiodine is not going to be needed when the patient is pregnant.

It is also based upon the fact that the ICRP has recommended enough radioiodine be cleared to ensure that the unborn child does not receive a dose in excess of 1 mGy unless, it is medically necessary for the health of the mother. There are occasional circumstances in which ^{32}P, ^{89}Sr, or ^{131}I meta-iodobenzylguanidine are used for therapy. In order to keep the dose to the fetus below 1 mGy, pregnancy should be avoided for 3, 24 and 3 months respectively, for the above mentioned isotopes.

Radiotherapy

The patient diagnosed as cervical cancer may be treated with radiotherapy with external beams and brachytherapy. If she is found to be pregnant, unfortunately, it is likely that pregnancy will be terminated. Carcinoma of the cervix is the most common malignancy associated with pregnancy. Cervical cancer complicates about one out of 1,250–2,200 pregnancies. This rate, however, varies significantly across countries. Cervical cancer is often treated by surgery or radiotherapy and the absorbed doses required with both forms of radiotherapy will cause termination of pregnancy. If the tumor is infiltrative and is diagnosed late in pregnancy, an alternative is to delay treatment until the baby can be safely delivered.

Regardless of protective measures, radiotherapy involving the pelvis of a pregnant female almost always results in severe consequences for the fetus, most likely fetal death.

In the case of breast cancer patient who had radiotherapy treatment, further pregnancy can be avoided. The wait can be substantial and needs to be discussed with her radiation oncologist. Most radiation oncologists advise their patients not to become pregnant for 1–2 years after completion of therapy. This is not primarily related to concerns about potential radiation effects, but rather to considerations about the risk of relapse of the tumor that would require more radiation, surgery or chemotherapy.

BIBLIOGRAPHY

1. Charles A Kelsey, Philip H Heintz, Daniel J Sandoval, et al. Radiation Biology of Medical Imaging. Hoboken, New Jersey: Wiley Blackwell; 2014.
2. Eric J Hall, Amato J Giaccia. Radiobiology for the Radiologist, 7th edition. Philadelphia, USA: Wolters Kluwer/Lippincott Williams & Wilkins; 2012.
3. International Atomic Energy Agency. Radiation biology: A handbook for teachers and students, Training course series 42. Vienna: International Atomic Energy Agency; 2010.
4. International Atomic Energy Agency. Safety report series No. 2: Vienna: International Atomic Energy Agency; 1988.
5. International Commission on Radiological Protection. Pregnancy and Medical Radiation. ICRP Publication 84. New York: Elsevier; 2000.
6. International Commission on Radiological Protection. Recommendations of the International Commission on Radiological Protection. ICRP Report 103. New York: Elsevier; 2007.
7. Jerrold T Bushberg, Anthony Seibert J, Edwin M Leidholdt JR, et al. The Essential Physics of Medical Imaging, 3rd edition. Philadelphia, USA: Wolters Kluwer/Lippincott Williams & Wilkins; 2012.

Biological Basis of External Beam Radiotherapy

CHAPTER 8

EVALUATION OF RADIOTHERAPY

Radiotherapy is one of the most effective methods for treatment of cancer. About two thirds of cancer patients are treated by radiotherapy either by alone or combination with chemotherapy and surgery. X-ray is the first radiation used in radiotherapy, after the discovery of X-rays by Wilhelm Conrad Roentgen, the German physicist in 1895. Australian surgeon Leopold Freund has demonstrated the first therapeutic use of X-rays, the disappearance of a hairy mole by treating X-rays in 1896 before the Vienna Medical Society. Since then, X-ray is the major source of radiotherapy, especially for the treatment of cancer. From 1951 onwards, Cobalt-60 teletherapy machines came into clinical use by replacing the low energy, kilo voltage X-ray units. The use of linear accelerators with high-energy X-rays came into clinical use in 1971, though the concept was known much earlier. Thus, megavoltage radiations started their impact in radiotherapy and still it continues in various forms of technology.

Usually radiotherapy is carried out in two forms, namely external beam radiotherapy and brachytherapy. Better understanding of the biology of tumors and normal tissues with respect to their response to radiation is very much essential for radiotherapy. The main objective of radiotherapy is to give maximum dose to the tumor and minimum dose to the critical normal tissues. The biological effect is influenced by dose rate and fractionation of the radiation. For a given dose, an increase in dose rate leads to an increase in biological effects. Hence, higher dose rates are advantageous for tumor control, but the disadvantage is the risk of normal tissue complications. The dose rate effect plays an important role in brachytherapy.

A radiation dose delivered in multiple fractions give lesser normal tissue complications, than a single acute dose. The fractionation came into radiotherapy due to low output of X-ray machines, since several hours were required to deliver the full dose. Emil Grubbe in Chicago has administered a course of radiotherapy with 18 fractions of daily 1 hour treatment for a breast cancer patient. Probably, this is the first fractionated radiotherapy treatment in the world. However, the clinical importance of fractionation was realized only after the experiments conducted by Regaud and Henri Coutard (1932) from the Curie institute, Paris. Regaud sterilized the rams by irradiation of their testis without excessive damage to the skin of the scrotum by fractionation. If it was given as single dose, the skin damage would have been unacceptable. He postulated that testis is a rapidly dividing cell and it is good model to study cancer. Further studies of Coutard had established fractionation as a standard clinical practice. However, the rational for fractionation is explained by radiobiology in recent times. The basic radiobiological reasons for dose rate and fractionation effects are best understood in terms of repair, reoxygenation, repopulation and redistribution, which are called 4R's of radiotherapy.

EFFECT OF DOSE FRACTIONATION ON CELL SURVIVAL

Fractionated radiation treatment over a period of weeks gives better therapeutic ratio for most of the tumors than giving a single treatment. Therapeutic ratio is the ratio of tumor control probability (TCP) and normal tissue complication probability (NTCP) for a given radiation dose. The favorable therapeutic ratio is easily achieved in radiosensitive cancer, e.g. seminoma of the testis. However, the majority of cancer in clinical practice is carcinoma, which has limited radiosensitivity. Therefore, new strategies are tried to improve the desired therapeutic ratio, especially in carcinoma. Altered radiation dose fractionation schedule is one such strategy in radiotherapy. When radiation treatment is fractionated it is found that a much greater total dose is required to achieve a given level of biological damage. This reveals that recovery of radiation damage takes place between fractionation. The recovery is the process that includes both repair of damage and cell growth during the fractionation of treatment.

Survival curves of mammalian cells exposed to X- or gamma (γ)-rays exhibit a broad shoulder at low doses, followed by an exponential region. At low doses, the radiation cause sublethal damage (SLD), which gets repaired compared to high-dose region. Thus, the initial shoulder reflects the magnitude of SLD. Recovery of SLD can be demonstrated by exposing the cells to a fractionation of dose and cell survival is studied.

Ability of the cell to recover from the damage has been demonstrated by Elkin and Sutton by split course experiments. A given total dose when delivered as a single dose was found to be more effective compared to the same dose delivered in two or more fractions, separated by a time interval. Result of split course experiments observed in Chinese hamster cells is shown in Figure 8.1A. There is a steep increase in survival of cells with the increase in time interval between two doses. The rise in survival during the first 2 hours is a result of the repair of SLD. Thereafter, the survival decreases, reaches a dip at around 6 hours and once again increases. The repair is a combination of three important processes that occurs simultaneously, i.e. repair, redistribution and repopulation.

First, the cell repairs the SLD; secondly, progression takes place along with synchrony. That is, the cells redistribute through the cell cycle during the interval between split courses and get synchronized. Thirdly, cell division takes place resulting in an increased number of surviving cells. As a result of the first dose fractionation, most of the cells in the sensitive region are killed; those which survive are mostly the ones in the radioresistant S phase of the cell cycle. The variation in the cell survival seen beyond 2 hours interval (refer Fig. 8.1A) results from the progression of the cells into other phases of the cell cycle varying in radiosensitivity, during the incubation interval. This is followed by repopulation process.

Recovery from SLD can be demonstrated by exposing the cells to a single dose (dotted line) and after 2 hours of incubation under suitable conditions. The result of the experiment is shown in Figure 8.1B, which is called split course survival. The results indicate that the cells incubated for a period of 2 hours after a dose, respond to a series of doses exactly like the unirradiated cells (continuous line). Reappearance of the shoulder confirms that the cells have recovered from the SLD during the interval period. This has been confirmed in different cell systems both in vitro and in vivo for several biological end points.

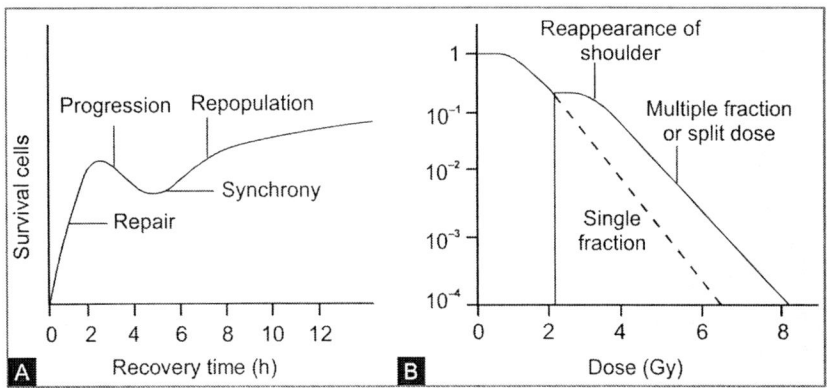

FIGURES 8.1A and B: Dose fractionation and cell survival. **A.** Recovery of irradiated cells overtime; **B.** Survival curve for single-dose and split-dose fractions.

The recovery from SLD during fractionation decreases with increasing LET. The recovery effect of dose fractionation for gamma rays is lesser for neutrons and totally absent for alpha rays. Recovery from SLD is also absent under acutely hypoxic conditions, but occurs under moderately hypoxic conditions. Mammalian cells exhibit both broad and narrow shoulder. A wide shoulder corresponds to small α/β ratio. The quadratic component 'β' is responsible for the sparing effect of split dose.

Dose fractionation technique has been routinely used in radiotherapy. Fractionation enables the normal tissue to recover between the fractions, and hence, reduces the damage to the normal tissue. Ability of the normal tissue to repair the radiation damage better than the tumor tissue forms one of the important bases of fractionated radiotherapy. Besides this, fractionation also helps the reoxygenation of hypoxic cells and redistribution of the cells in the sensitive phases of the cell cycle. It is likely that hypoxic cells, which survive may show a lesser ability to recover from SLD compared to well-oxygenated tissue. Some of the tumor cells, which show very wide initial shoulder, reflect their ability to accumulate and repair a large extent of SLD. This could possibly be the reason for the failure of radiotherapy in certain patients.

EFFECT OF DOSE RATE ON CELL SURVIVAL

Survival curves of cells irradiated at low-dose rates (cGy per hour) show much lesser steepness (less biological effects) compared to cells irradiated at higher dose rate (cGy per minute). Radiation exposure with low-dose rate and long duration reduce the biological effects relatively. This implies that the cell line is resistant at low-dose rate irradiation due to division of cells during irradiation. Variation in response to treatment with dose rate is referred to as dose rate effect and is found to be most important in the dose rate range of 1–1,000 mGy/min, using in vitro Chinese hamster cells exposed with Cobalt-60 gamma rays (Fig. 8.2).

As the dose rate is reduced from 1,000 mGy/min, then the slope of the survival curve become shallower and shoulder tends to disappear. This can be explained as follows: When the dose rate is higher, the curve has a significant initial shoulder. As the dose rate is reduced, the SLD is repaired, the curve becomes shallow and the shoulder decreases. If the dose rate is decreased still further, more and more SLD repair takes place and

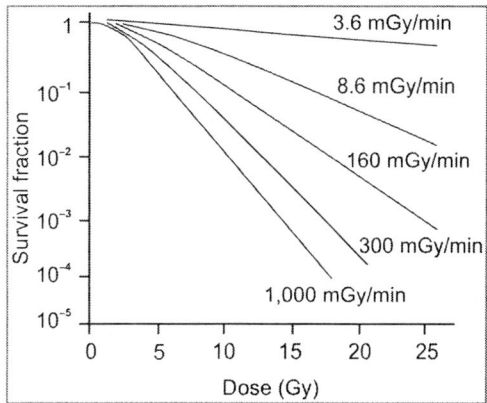

FIGURE 8.2: Dose rate effects on Chinese hamster cells, grown in vitro, exposed to Cobalt-60 gamma rays.

the shoulder tends to disappear. This means that the cell has reached a point, where all SLD is repaired. If the dose is fractionated for a given dose rate, the SLD repair process gets accelerated still further.

However, in some cell lines, further lowering of dose rate permits the cells to progress through the cell cycle and accumulate in G2 phase. Since it is a radiosensitive phase, the survival curve becomes steeper again with reappearance of shoulder. This is called inverse dose rate effect. Further reduction of dose rate allows cell to pass through G2 phase and starts dividing. The radiation response decreases further and curve becomes shallow again. The inverse dose rate effect has vital role in brachytherapy.

The dose rate effect is much less for high-LET ionizing radiations compared to low-LET X- and γ-radiations. This may be due to the differences in the nature of the damage induced by these radiations. High-LET radiations usually cause double strand breaks in DNA, due to dense ionization. Hence, damage induced by the high-LET radiations are complicated to repair compared to low-LET X- and γ-rays with low-dose rates.

Dose rate effect has very important implications in radiotherapy. For effective cell killing, the dose rate should be above a critical value at which the cell population is rapidly depleted. Under critical dose rate, the cells are arrested in mitosis and die while attempting to division. Human tumors require a dose rate of 0.3–0.5 Gy per hour in brachytherapy. Induction of genetic effect in vivo is also found to be less (3–10 times) in low-dose rate compared to high-dose rate. For most mammalian cell systems, the dose rate effect is very rare, beyond 200 cGy min^{-1}.

RATIONALE FOR FRACTIONATION IN RADIOTHERAPY

The justification of using fractionated radiotherapy can be fully understood only by radiobiological basis. Radiobiology emphasize that there are 4 Rs of radiotherapy; namely, repair, reoxygenation, repopulation and redistribution, which are very important in fractionation.

Repair

Repair relates to the ability of a cell to recover from damage to its vital genetic structure before further damage. It is otherwise called intracellular recovery. The repair takes place both in normal and tumor cells. Late reacting normal tissues tend to be more capable of repairing than the cells in tumor and early reacting normal tissues. The repair mechanism is important at low doses and was referred as Elkind type of recovery, which is highly dependent on cell type. The degree of repair is strongly dependent on interfraction time, number of fractions, dose per fraction and dose rate, and is independent of total dose and overall treatment time. The cell survival curve for late reacting normal tissues tend to be curvier than that of tumor cells, due to their low α/β values (Fig. 8.3).

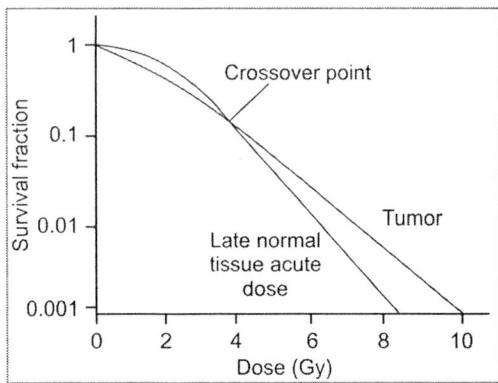

FIGURE 8.3: Survival curves for tumor and late reacting normal tissues. The normal tissue curve is curvier than tumor.

The normal and tumor cell curves crossover at doses of 3–5 Gy. At low doses (before the crossover point), more tumor cells are killed than late reacting normal cells. Above the crossover point, more late reacting cells are killed than the tumor cells. Thus, if the dose is > 5 Gy, it will affect the normal tissues, even though tumor needs a dose > 5 Gy. This can be overcome by two ways:

1. Delivering the total dose in multiple factions.
2. To deliver much higher dose to the tumor than the normal tissue as in stereotactic radiosurgery in a single fractionation.

The multiple fractionation protocol is designed in such a way that there should be sufficient time interval between fractions, which is sufficient to repair completely the SLD. That is, the cells damaged during the first fraction will be repaired before the second fraction and so on. As a result, the surviving fraction after each treatment is identical. Hence, the shape of the survival curve is also same that will repeat after each fraction. Then, the dose per fractionation must be below the crossover point, so that the tumor cells get more damaged.

If the dose is fractionated, the survival curve of tumor and normal tissue gets well separated. This suggests that there is repair in between each fraction that depends on the dose per fraction. The maximum separation of the curve refers to complete repair, which is possible only at very low dose per fraction with infinite number of fractions. In clinical practice, it is not possible, since repopulation may counteract the repair effects. However, for optimal repair, if the dose per fraction size is before the crossover point, the tumor survival curve is below the normal tissue curve, suggesting higher damage to the tumor (Fig. 8.4). Alternatively, if the dose per fraction size is above the crossover point, the normal tissue survival curve is below the tumor curve, suggesting higher damage to the normal tissue, which is not accepted (Fig. 8.5). Hence, the optimal dose per fraction is somewhere before the crossover point.

In the first survival curve (refer Fig. 8.3), the maximum separation occurs around a dose per fraction of 1.9 Gy, before the crossover point. This is exactly equal to 50% of the crossover point dose, which is about 3.8 Gy. Since the crossover point dose vary from 3 to 5 Gy, it is logical to have optimal dose per fraction in 1.5–2.5 Gy range. If the 'α' and 'β' values of the specific tissues are known, then it is possible to design individual patient specific fractionated treatment schedules with optimal dose per fraction.

However, it is interesting to know how high-dose fraction or enhanced high total doses are delivered in IMRT or 3D-CRT or

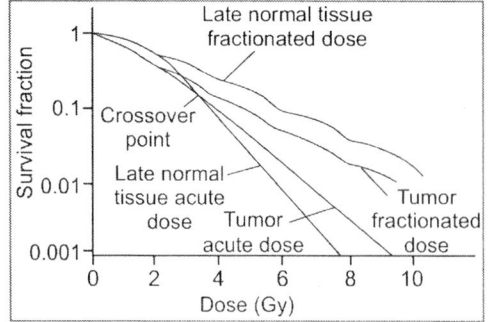

FIGURE 8.4: Survival curve for fractionated dose with dose per fraction that is below the crossover point, predicting more damage to tumor cells.

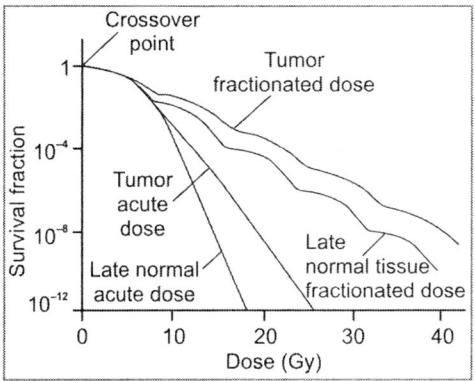

FIGURE 8.5: Survival curve for fractionated dose with dose per fraction that is above the crossover point, predicting more damage to late reacting normal tissues.

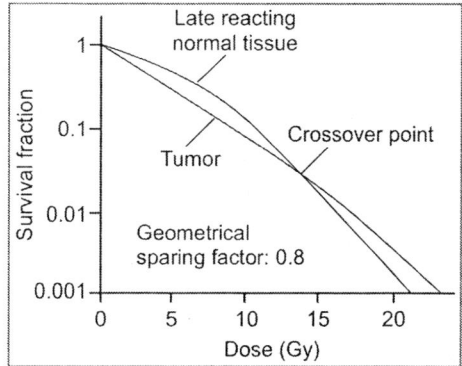

FIGURE 8.6: Crossover point movement with geometrical sparing factor of 0.8, suitable for single high-dose per fractionation.

stereotactic radiosurgery or cyber knife irradiation. In these treatments, the effective dose to normal tissues is kept well below the tumor by offering suitable geometric sparing and the crossover point is moved to higher doses. The geometrical sparing factor is the ratio of the effective dose in normal tissues to effective dose in tumor. Even for a minimal geometrical sparing factor of 0.8, the crossover point can be moved to a higher dose and widens the therapeutic window (Fig. 8.6). For example, the crossover point can be moved from 3.8 to 14 Gy in such treatments. Hence, 50% of the crossover point dose that is 7 Gy dose per fraction size, is possible in such situations. This is similar to that of HDR brachytherapy treatments in carcinoma cervix, where the geometrical sparing is inherent.

In the case of stereotactic radiotherapy, doses of the order of 20 Gy are used. This is under the assumption that there is geometrical sparing of 0.6. However, this is true only for small tumors, whereas for large tumor, it is difficult to achieve a geometric factor of 0.6. The solution to this issue is the conventional fractionation protocol.

Reoxygenation

Oxygen is the most powerful radiation sensitizer. It is well-known that the recovery of SLD is oxygen dependent and severe hypoxic condition reduces the rate of repair. Metabolic activity is required for repair, and hence, hypoxic cells are slow to repair sublethal injury. Normal cells are considered well oxygenated, whereas tumor cells are hypoxic or even anoxic. Therefore, normal cells have greater recovery capacity from SLD than tumor cells.

The cells deprived of oxygen (hypoxic cells) are relatively resistant to radiation and require about 3 times as much dose to destroy them. This will certainly exceed the normal tissue tolerance. There is an evidence that tumor contains a significant proportion of hypoxic cells. During fractionated radiotherapy, the oxygenated cells are killed due to their radiosensitivity. As the fraction increases, the proportion of hypoxic cells should increase as more and more oxygenated cells are killed. However, the experiments have predicted that the fraction of hypoxic cells decreases compared to pre irradiation, over fractionation. This suggests that during fractionation, hypoxic cells get oxygenated due to tumor shrinkage, as they move closure to the vasculature. Those cells, which are hypoxic in the early fraction become oxygenated and may be killed in the subsequent fractions, which is known as reoxygenation.

The time interval between fractions should be sufficient to offer full reoxygenation to hypoxic cells. In fractionated radiotherapy, a 24-hour interval is sufficient to treat the tumor effectively without exceeding normal tissue tolerances. Thus, fractionation is essential for tumors having hypoxic cells. However, there is some evidence that fractionation does not always ensure sufficient reoxygenation to overcome hypoxic cell radioresistance. Hence, even with fractionated radiotherapy not all cancers are curable. The mechanism of reoxygenation is found to vary considerably from tumor to tumor. The preirradiation hypoxic level and duration of reoxygenation of all tumors are not well understood. It was suggested that the optimum interval required for reoxygenation was shorter for small dose per fraction. Therefore, short treatment regimens with large dose per fraction would result in less reoxygenation than a longer treatment regimen with small dose per fraction, which is given in several times per day.

Repopulation

When a body tissue is exposed to ionizing radiation, the tissue responds to the radiation injury by repopulation or regeneration of damaged tissues by the process of cell proliferation. It mainly differs in its rate of repopulation from repair process. While repair occurs immediately within few hours of radiation exposure, repopulation is a long-term process, which may continue for several weeks even after the completion of treatment. The degree of repopulation is dependent on overall treatment time, tissue type and important in rapidly proliferating tissues.

Repopulation represents extracellular recovery. Repopulation occurs both in normal cells and tumor cells, but the rate of repopulation is better in tumor cells than in normal cells. Repopulation takes time to occur and will depend largely on the proliferative properties of the tissue. Most of the late normal tissue damages are believed to be the result of radiation damage in slowly proliferating cells. This may be the reason for lesser influence of overall treatment time on late normal tissue reactions.

All cancer tissue consists of fast dividing cells and usually divides in much faster rate than those of late reacting normal tissues. Hence, repopulation is very high in tumors due to accelerated growth rate that might occur after irradiation. During a course of radiotherapy, there will be more repopulation of cancer cells than late reacting normal cells. Longer the duration of radiotherapy, higher is the repopulation of tumor. Thus, it is difficult to control tumor without exceeding the normal tissue tolerance. Repopulation of cancer cells gets accelerated after first 2 weeks of treatment. Studies have shown that, in some cases, repopulation occurs even after 2-3 weeks of fractionated treatment.

Hence, the schedules of treatment need to be as short as 2-4 weeks. The normal tissue cells have to repopulate during the course of radiotherapy, to avoid exceeding acute tolerance. Hence, the optimal duration of fractionation has to satisfy two things:
1. Acutely responding normal tissues need to repopulate during fractionation.
2. Not to allow much time for excessive repopulation of tumor cells.

If more number of fractions is given per day, better the tumor control. The success of treating rapidly growing tumors mainly lies on controlling repopulation.

Redistribution

The sensitivity of the cells vary with the phase of the cycle. The cells irradiated during the mitotic phase (M) and late G2 of the cell cycle are more sensitive, and at late S phase are more resistant. When an asynchronous population of cell is irradiated, those in the resistant phase of the cycle will be most likely to survive and a certain degree of synchrony is achieved in the cell cycle. The distribution of cells in the cell cycle is called redistribution. By suitable adjustment of timing of succeeding fractions, it is possible to fix the cells in their most wanted phase of the cell cycle. The therapeutic ratio could be maximized by optimal choice of interfraction interval. This interval depends upon tumor cell type, type of normal cells around the tumor and the degree of oxygenation.

The cells in the M phase, the survival curve is linear indicating that there is minimal repair. In the late S phase, the curve exhibits greater shoulder corresponding to maximal repair. After a single fraction of radiation dose, there is an abundance of surviving cells in late S phase. The success of second dose of delivery will depend on how far these surviving cells have travelled around the cell cycle. If they have reached the M phase during the second fraction, they will be more sensitive and are effectively killed. Thus, the radiation-induced partial synchronization of cells is achieved and is known as redistribution. Redistribution is beneficial in fractionated radiotherapy in which the cancer cells are caught in mitosis. It could be detrimental if normal cells are caught in mitosis. If the inter fraction interval is adjusted well, maximum benefit is obtained from redistribution. However, in clinical practice this is difficult to follow.

Human tumors did not have a uniform cell cycle. Since, tumor cells are proliferating at faster rate than normal cells, the sensitizing effect on redistribution would be greater on tumor cells. This effect is greatest, when small dose per fraction is used. Since, non-proliferating tissues limit the dose in radiotherapy and are not be sensitized by redistribution, the more the dose is fractionated and lower the dose per fraction, the better would be the tumor control. The repopulation of the tumor could offset the tumor cure, if the overall treatment time is longer. However, it can be overcome by delivering multiple fractions per day, allowing redistribution and hitting the cells in their sensitive phase of the cell cycle.

Overall Effects of 4 Rs

1. Repair is the major reason for the dose rate effect and it is important at low doses and dose rates. It favors late responding normal tissues relative to tumors and provides a rationale for fractionation in external beam radiotherapy.
2. Repopulation, on the other hand, fits tumors more than late reacting normal tissues, especially tumors with short potential doubling time. Hence, it is important to control the overall treatment time.
3. Redistribution is of little consequence in radiotherapy, as it would be difficult to utilize its advantage.
4. Reoxygenation is probably important for many tumors. Experiments show that it varies from one type of tumor to another.

DOSE-RESPONSE CURVES

The dose response refers to the relation between radiation dose and the percentage of control of biological response. Holthusen (1936) illustrated graphically the tumor regression and tissue damage as a function of dose. He was the first to give theoretical analysis of dose-response relationship. As the dose increases, the biological response also increases in severity and frequency or both. The dose-response curve has sigmoid in shape and this is true for both normal tissue reactions and local cancer control (Figs 8.7A and B). The left curve refers to the probability percentage of tumor control and the right one refers the probability percentage of incidence of normal tissue damage (refer Fig. 8.7A). He noted that the resemblance between the curves reflect the variation of clinical response of individual patients. The response is zero at zero doses and tends to be 100% at very large doses. It also emphasizes that the normal tissue response occurs at higher doses than tumor response. However, higher tumor response always involves some normal tissue response. There are several factors that modify the dose response of the system. In Figure 8.7A, tumor control is maximum and the normal tissue damage is minimum for a particular dose, which is favorable to radiotherapy. On the other hand, in Figure 8.7B the situation is not favorable for radiotherapy.

The main aim of radiotherapy is to achieve high tumor control probability (TCP) and a low normal tissue complication probability (NTCP). Both TCP and NTCP increases with dose and their relationship is well described by the dose-response curve, which is sigmoid in nature. The therapeutic ratio or therapeutic index is the parameter used to quantify their relationship. It is defined as the ratio of damage to tumor cells to damage to the normal cells for a same given dose:

$$\text{Therapeutic ratio} = \frac{\text{Damage to tumor cells}}{\text{Damage to normal cells}} \quad \text{-----(1)}$$

It is difficult to assess the damage of cells, since damage is not linear with dose. Then, the easy way of defining the therapeutic index is the ratio of the dose for a given biological end point:

$$\text{Therapeutic ratio} = \frac{\text{Dose to normal cells}}{\text{Dose to the tumor cells}} \quad \text{-----(2)}$$

Therapeutic ratio is a measure by which one can realize and achieve the objective of radiotherapy. Fractionation scheme can be designed to achieve a given therapeutic ratio in planning. A typical example is hyperfractionation, which reduces the normal tissue complications and increases the therapeutic ratio. Similarly, administration of radiosensitizers will enhance tumor control without harming the normal tissues.

These hypothetical dose-response curves are obtained by using the logistic model for which the probability (P) of given effect is given as follows:

$$P = \frac{1}{1 + [D_{50}/D]^k} \quad \text{-----(3)}$$

where,

D is the total dose and D_{50} is the dose that has probability of causing the effect in 50%, and k is the tissue specific parameter that regulates the slope of the dose-response curve at

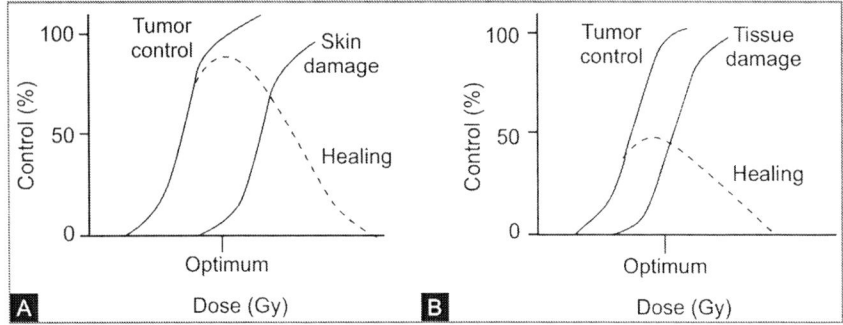

FIGURES 8.7A and B: Dose-response curve for both tumor and normal tissue. A. Increasing percentage of tumor control and minimal normal tissue damage, favorable condition; B. Increased tumor control and increased normal damage, unfavorable condition.

dose $D = D_{50}$. The biological effective dose is given by the linear quadratic model as:

$$BED = D\left(1 + \frac{d}{\alpha/\beta}\right) \quad \text{-----(4)}$$

Here, the repopulation is ignored. For 50% of tumor control probability, the above equation is rewritten as:

$$BED_{50} = D_{50}\left(1 + \frac{d}{\alpha/\beta}\right) \quad \text{-----(5)}$$

By dividing the above two equations (4 and 5), one can get:

$$\frac{D_{50}}{D} = \frac{BED_{50}}{BED} \quad \text{-----(6)}$$

Substituting the above value in the logistic equation (3):

$$P = \frac{1}{1 + [BED_{50}/BED]^k} \quad \text{-----(7)}$$

In order to obtain suitable values for BED_{50} and k, the tumor response curve is normalized. The normalization is done at 80% local control by a total dose of 50 Gy, in 2 Gy per fraction (Fig. 8.8). If the dose is reduced to 40 Gy from 50 Gy, local control probability drops to 50%. The resulting values of BED_{50} and k are 48 and 6.2 respectively for a α/β value of 10 Gy. Similarly, the late responding normal tissue curve is normalized with severe injury of 3% for 50 Gy and rising to 50% injury for 70 Gy respectively. The resulting values of BED_{50} and k are 126 and 10.3 respectively for a α/β value of 2.5 Gy. Using the above values, the dose-response curve is constructed for a daily 2 Gy per fraction (refer Fig. 8.8).

The nature of fractionation scheme decides the percentage of tumor control and normal tissue complications. Generally, fractionation schemes are designed with little flexibility over dose. A dose that gives ±10% of tumor control or ±5% probability of late normal tissue complications is clinically acceptable. If we assume that a 10% reduction

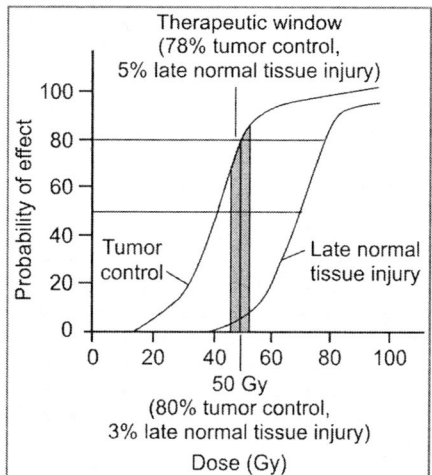

FIGURE 8.8: Dose-response curve illustrating the tumor control verses normal tissue injury with 2 Gy per fraction. The therapeutic window will decrease if the dose per fraction increases.

in tumor control is clinically acceptable, then the tumor control probability becomes 72%, instead of 80%. Similarly, if the rise of late tissue complications is clinically acceptable from 3 to 5%, then the total dose has a flexible region called therapeutic window. This acceptable variation of tumor control (72%) and risk of late reacting normal tissue injury (5%) is usually termed as therapeutic window. Within this window various fractionation schemes can be designed to achieve desired goal.

If the window is narrow, it is difficult to prescribe a safe and effective dose. If the window is wider, it is easy design a suitable fractionation scheme to the patient. On the other hand, if the window is negative, the above concept itself breaks down, either the local control is too low or the normal tissues has exceeded its tolerance limit. Therapeutic window width is influenced by the magnitude of dose per fraction. Instead of 2 Gy per fraction, if we use 3 Gy per fraction, the window not only shrinks but also moves toward left (Figs 8.9A and B).

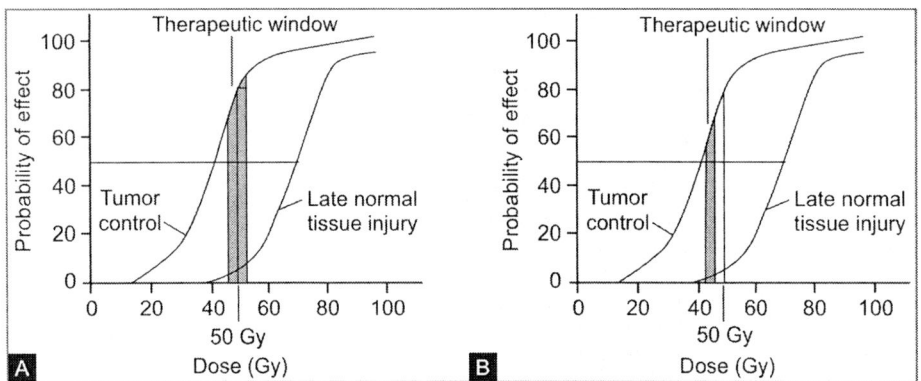

FIGURES 8.9A and B: Dose-response curves. **A.** 2 Gy per fraction has wider therapeutic window; **B.** 3 Gy per fraction has narrow window.

In general, low dose per fraction provides wider window, whereas higher dose per fraction provides a narrow window. It is found that a dose per fraction of 2–3 Gy gives a clinically acceptable therapeutic window. Treatment schedules within the therapeutic window often offer sufficient damage of tumor cells without exceeding the normal tissue tolerance. If the dose per fraction is > 3 Gy, therapeutic window becomes negative and gives unacceptable risk of excessive normal tissue injury or low tumor control. However, there is a geometrical sparing factor in all modern radiotherapy treatments, especially in IMRT and 3DCRT, etc. The earlier discussion has not taken into account the geometrical sparing effect. If a geometrical sparing of about 20%, which is easily achievable in IMRT is considered, there is provision to increase the dose per fraction. To understand this fact, one should draw the relation between therapeutic window and dose per fraction (Figs 8.10A and B). In Figure 8.10A, the clinically acceptable dose per fraction is shown between 2 and 3 Gy. Above 3 Gy per fraction, it is not possible to design a fractionation scheme, since it is more toxic. The Figure 8.10B shows the fractionation schemes without geometrical sparing and with 20% geometrical sparing. This not only reduces the high-risk region but also provides an opportunity to increase the dose per fraction up to 4.5 Gy per fraction.

FRACTIONATION STRATEGIES

In order to maximize the radiobiological advantages, the fractionation schemes in clinical radiotherapy are employed as:
- Conventional fractionation
- Hyperfractionation
- Accelerated fractionation
- Accelerated hyperfractionation
- Dynamic fractionation
- Hypofractionation.

Conventional Fractionation

The conventional fraction scheme consists of 1.8–2.2 Gy per fraction, delivered in 5 fractions per week. It is in use for more than 50 years and is convenient, efficient and effective. One can go up a total dose of 60–81 Gy over a period of 5–6 weeks. This scheme facilitate no weekend treatment yet deliver high doses to the tumor without exceeding both acute and chronic normal tissue tolerance. The special feature of the scheme is that the tumoricidal and the normal tissue tolerance doses are well documented by experience.

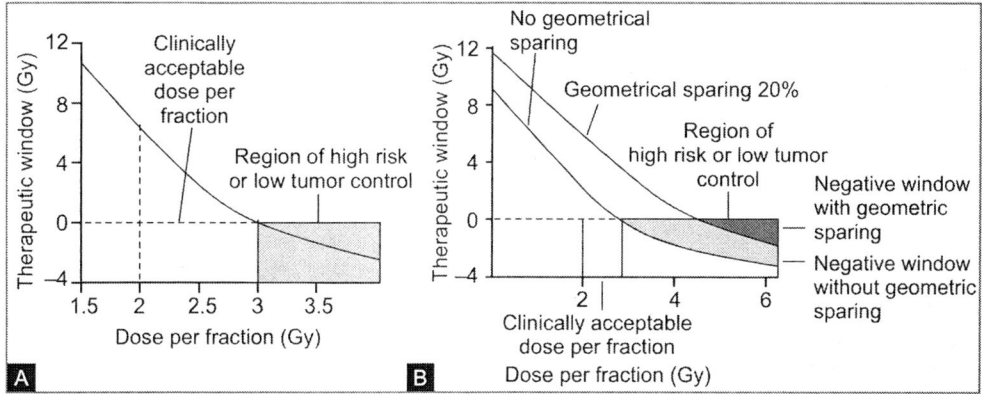

FIGURES 8.10A and B: Relation between therapeutic window width and dose per fraction. **A.** Conventional fractionation without geometrical sparing; **B.** Higher dose per fraction of 4.5 Gy per fraction with geometrical sparing of 20%.

Hyperfractionation

The hyperfractionation is one in which more than one fraction is delivered each day, by keeping the overall treatment time same as the conventional fractionation. The aim is to increase TCP without increasing NTCP. This is achieved by providing sufficient time between the first and the second daily fraction, so that complete repair of SLD does occur, before the delivery of second daily fraction. Animal study reveals that the half-time repair is 1 hour, but there is wider variation of time in humans. Hence, at least 6 hours between daily fractions are required.

The typical dose schedule is 1.2–1.3 Gy per fraction, delivered in 2 fractions per day. This will permit the enhancement total dose by 20–30%, due to the efficient repair at low dose per fraction. The rationale for hyperfractionation is that it maximizes the difference between the repair capacity of late reacting normal tissue and tumor. It is believed that the radioresistant tumor cells that survive during the daily first fraction will be available at a more sensitive point in the cell cycle during the second daily fraction, due to redistribution.

If conventional fraction fails to offer expected clinical results, it means that dose per fraction has not maximized the difference in repair capacity of normal and tumor cells. These types of tumors are more suitable candidate for hyperfractionation. Such hyperfractionation should not be over protracted, so that there is a risk of tumor cell repopulation. A hyperfractionation of > 1.3 Gy per fraction, delivered in 2 fractions per day may exceed the acute tolerance of the normal tissue. On the other hand, a dose fraction of < 1.2 Gy, require 3 fractions per day to keep the overall treatment time same and it is less dependence of hypoxic cells. Hence, mostly 1.2–1.3 Gy per fraction is practiced in hospitals. The symbol # is commonly used to mention fraction.

Accelerated Fractionation

Accelerated fractionation is designed to overcome repopulation of tumor cell during radiotherapy. The pure accelerated fractionated scheme requires that the duration of the radiation treatment is shortened without changing the fraction size and the total dose. This can be done either using twice daily fractions during some of the week days or once daily fraction for 6 days instead of 5 days. However, it provides only modest acceleration. Generally, intense accelerated fractionation results in acute normal tissue effects and requires interruption in the treatment, in addition to increased NTCP for late effects.

In practice, in accelerated fractionation, treatment is offered over 6–7 days per week, instead of 5 days a week, by keeping the dose per fraction as same as conventional fraction. This will offer intermediate acceleration that is sufficient in most clinical situations. This is suitable for rapidly growing tumors with short potential doubling times of viable and cycling tumor cell. It is more suitable for tumors that exhibit accelerated repopulation.

The level of acceleration can be increased either with 2.5 Gy per fraction or 1.4 Gy–1.6 Gy per fraction, delivered in 2 fractions per day. However, the former lose the advantage of late responding normal tissue, exhibiting late reactions. On the other hand, the later scheme may exceed the acute normal tissue tolerance. However, this can be overcome by accelerated hyperfractionation protocol.

Accelerated Hyperfractionation

Accelerated hyperfractionation combines the advantage of hyperfractionation and accelerated fractionation. This will improve TCP at the same time it ensures that the acute normal tissue effects do not increase NTCP. It is possible to exploit the difference in repair capacity between the late reacting normal tissues and tumor by hyperfractionation. At the same time, it is possible to accelerate the course of therapy by treating over 6–7 days with multiple fractions per day. The overall treatment time is much reduced. Such a fractionation scheme is called accelerated hyperfractionation. This can be done in three ways:
1. Continuous hyperfractionated accelerated scheme.
2. Split course accelerated fractionation scheme.
3. Concomitant boost accelerated fractionation scheme.

The continuous hyperfractionated accelerated radiotherapy (CHART) protocol of Mount Vernon Hospital, London, advocates a dose fraction of 1.5 Gy, 3 fractions per day, 6 hours apart, 7 days a week, making the total dose of 54 Gy delivered in 36 fractions over 12 successive days, including weekends. The first 28 treatments are offered with large fields covering both primary and nodal regions. The remaining 8 treatment is offered as boost covering only the primary tumor region. However, patients often developed with acute reactions and trauma with post-treatment hospitalization. Delivering 3 fractions per day, 6 hours apart and weekend treatment cause some difficulty to the hospital staffs.

Hence, the CHART is modified to overcome the above inconvenient, it is called CHARTWEL. This scheme recommends a dose of 54 Gy at 1.5 Gy per fraction at 3 fractions daily, is delivered over a total of 16 days, without weekend treatment. Overall, it focuses on two issues; one is to reduce the duration of the treatment. This reduces repopulation in the tumor, while allowing complete repair of SLD. Second is the reduction of total radiation dose to about 10–18%, to reduce the expected acute normal tissue effects.

The split course accelerated fractionation study (RTOG, 2000) recommends delivery of 67.2 Gy/42#/6.2 weeks treating 1.6 Gy 52# per day in 2 phases with 2 weeks break in between. In this split course, accelerated fractionation allows recovery of acute normal tissue effects, since there is 2 weeks break between phases. No improvement in local control and no worsening late normal tissue complications are seen. The concomitant boost accelerated hyperfractionated study (MD Anderson, 2001) involves 63 Gy/35#/5 weeks, treating 2#/day during last 2 weeks. No statistically significant difference in local control or late normal tissue complications is noticed.

Dynamic Fractionation

The other way of achieving accelerated hyperfractionation is dose escalation. It is possible to deliver a total dose of 76 Gy in 5 weeks as given below:

- Two fractions daily, 1.2 Gy per fraction, 6 hours apart, for the first 2 weeks, followed by
- Two fractions daily, 1.4 Gy per fraction, 6 hours apart, for the next 3rd and 4th week, and
- Two fractions daily with 1.6 Gy per fraction in the 5th week.

Apart from this, a single dose of 2 Gy is delivered on every Saturday. The acute reactions can be reduced suitably by reducing the field size. To control the acute late reactions, it is advised to reduce the field size gradually.

Hypofractionation

Hypofractionation schemes are aimed at curative intention. However, in the case of palliative conditions such a complicated and costly long-term treatment is not required. It is possible to design a short-term scheme with higher dose per fraction, which is called hypofractionation. For example, 10 fractions of 3 Gy or a single fraction of 10 Gy or any fractionation scheme in between them is possible over a period of 1–5 weeks. Of course, such hypofractionations are found to be useful in curative treatment also, e.g. stereotactic radiosurgery and HDR brachytherapy, etc. Hypofraction is suggested for tumors having lower α/β value than the late reacting normal tissues, e.g. prostate.

DOSE FRACTIONATION EFFECTS IN EXTERNAL BEAM TREATMENTS

Head and Neck Cancer with Radiotherapy

The α/β ratio of head and neck is higher than that of late normal tissue. Treatment of more, smaller dose fractions to a higher total dose is advantageous. However, prolongation of treatment time may be associated with loss of local control, 1–2% per day of prolongation (Dale RG et al, 2002; Selvin NJ et al, 1992). Hence, the treatment time must be kept as short as possible. Treatment time of < 20 days may bring much more improvements. The clinically observed local control rate for T1 cancer is about 85% (Saarilahti et al, 1998; Brouha XDR et al, 2000). In the case of advanced cancer, it is about 30–50% using conventional fractionation regimes (Stuschke M et al, 1999; Dische S et al, 1997; Fowler JF et al, 2000). Randomized trials of hyperfractionated regime have improved local control rates to 40–60% for intermediate to advanced cancers (FKK et al, 2000).

Trials of accelerated fractionation without reduction in total dose have improved local control rate in mixed stage of cancers. However, if the time is reduced too much, there is no significant improvement in local control (Horiot JC, 1997). For early stage cancers, a reduction in overall treatment time by increasing fraction size and using reduced total dose has shown improved local controls.

Head and Neck Cancer with Postsurgery Radiotherapy

The clinically observed local control rate for postoperative radiotherapy is about 70% (Muriel et al, Suwinki et al). No correlation is found between total dose and treatment time. Suwinki et al (2003) states that average dose intensity, i.e. total dose/total treatment time has influence on recurrence free survival. Muriel et al (2001) have shown that the total treatment time was significant prognostic factor. The absence of a significant dose-response function may be due to prescription bias and to the shallow slope of the dose-response curve for postoperative patients. Awwad HK et al (2002) have found improved 3-year local control (88 vs 57%) for patients treated with accelerated hyperfractionation (46.2 Gy/33#/12 days).

Breast Cancer

The fractionation regimes used in postoperative radiotherapy do not vary much. However, the boost, boost dose/fraction and type of boost (electron or photons) vary widely. Early stage breast cancer with breast conserving surgery and radiotherapy small variation of 5% with treatment regimes. However, most of the regime may yield 90% of local control.

It may be possible to decrease the total number of fractions and total dose, while maintain similar local controls (42.5 Gy/16# vs 50 Gy/25#). Randomized trials with reduced fractionation, but in a similar overall time of conventional fractionation showed that all regimes gave similar 10-year local control (42.9 Gy in 13# over 5 weeks vs 50 Gy in 25# over 5 week). Since the α/β ratio of breast tumor is small, use of hypofractionation may be advantageous; this is the current topic of interest.

BIBLIOGRAPHY

1. Charles A Kelsey, Philip H Heintz, Daniel J Sandoval, et al. Radiation Biology of Medical Imaging. Hoboken, New Jersey: Wiley Blackwell; 2014.
2. Colin G Orton. Fractionation: Radiobiologic principles and clinical practice. In: Faiz M Khan (Ed). Treatment Planning in Radiation Oncology, 2nd edition. Philadelphia, USA: Lippincott Williams & Wilkins; 2007.
3. Colin G Orton. Radiobiology. In: Subir Nag MD (Ed). High Dose Rate Brachytherapy. Armonk, NY: Futura Publications Company Inc; 1994.
4. Colin G Orton. Radiobiology. In: Subir Nag MD (Ed). Principles and Practice of Brachytherapy. Armonk, NY: Futura Publications Company Inc; 1997.
5. Eric J Hall, Amato J Giaccia. Radiobiology for the Radiologist, 7th edition. Philadelphia, USA: Wolters Kluwer/Lippincott Williams & Wilkins; 2012.
6. Johns HE, John R Cunningham. The Physics of Radiology, 4th edition. USA: Charles C Thomas; 1984.
7. Loredana G Marcu. Eva Benzak. Recent advances and Research Updates: Medical Physics and Radiobiology. Trivandrum, India: Researchman Publishers Pvt Ltd; 2011.
8. Mayles P, Nahum A, Rosenwald JC. Handbook of Radiotherapy Physics-Theory and Practice. New York: Taylor & Francis; 2007.
9. Sanjay S Supe. An overview of time dose fractionation models; NSD to BED. Pol J Med Phy Eng. 2006;12(4):165-201

Biological Basis of Brachytherapy

CHAPTER 9

BRACHYTHERAPY

Brachytherapy is a method of treatment in which sealed radioactive sources are used to deliver radiation at short distances, i.e. less than 5 cm. The sources may be placed on or near the tumor surface, or within the natural cavities or implanted directly into the tumor tissues. This is generally classified into:
- Intracavitary
- Interstitial implant
- Intraluminal
- Surface mold treatments.

Again, based on the dose rate, the brachytherapy is divided into:
- Low-dose rate (0.4–2 Gy/h)
- Medium-dose rate (2–12 Gy/h)
- High-dose rate (> 12 Gy/h).

Brachytherapy was started with low-dose rate (LDR) using radium (Ra)-226 sources. Radium may deposit in bone and the leakage of radon, a toxic gas, soluble in tissue, resulting in significant hazard to the workers. Hence, use of radium has been banned in hospitals. Nowadays, cobalt (Co)-60, iridium (Ir)-192, cesium (Cs)-137, iodine (I)-125, etc. are used as brachytherapy sources. Medium dose rate (MDR) is rarely practiced in clinical settings.

The source loading in brachytherapy treatment was started with preloading technique, which gives high-radiation exposure to the workers. In preloading technique, the sources are loaded into the applicator before its insertion into the patient. Hence, afterloading treatment technique came into existence. In this, the sources are loaded after the insertion of applicator into the patient, which reduced workers radiation exposure drastically. The afterloading can be done either by manual or remote loading techniques. High-dose rate (HDR)-brachytherapy remote afterloading machines are very popular today; and there are about 360 machines in India. Mostly, the above machines use either Ir-192 or Co-60 sources. The physical characteristics of few brachytherapy sources are given in Table 9.1.

Pierre Curie who has self-inflicted his forearm with radium burn is the beginning of radiobiology in brachytherapy (1901). The effect of radiation is the same both in teletherapy and brachytherapy. However, the spatial and temporal distribution of dose is different in brachytherapy. The parameters that influence such dose distributions are:
- Dose rate effect in LDR brachytherapy
- Fractionation in HDR brachytherapy.

HDR and higher dose per fraction cause more cell kill in normal tissues and unable to achieve local control. However, geometric sparing of normal tissue is possible in such brachytherapy. Hence, one can raise the dose rate and dose per fraction, and achieve local control without exceeding the normal tissue tolerance.

Table 9.1: Physical properties of brachytherapy sources

Source	Half-life	Photon energy (MeV)	HVL in lead (mm)	Exposure rate constant*(γ_δ) (Rcm²/mCi-h)
Radium-226	1,600 year	0.184–2.45 (average 0.83)	12	8.25
Cobalt-60	5.26 year	1.17, 1.33 (average 1.25)	11	13.07
Iridium-192	73.8 day	0.009–0.884 (average 0.397)	2.5	4.69
Cesium-137	30 year	0.662 (average)	5.5	3.26
Iodine-125	59.4 day	0.028 (average)	0.025	1.46
Palladium-103	17 day	0.021 (average)	0.008	1.48

LDR AND HDR BRACHYTHERAPY

In LDR brachytherapy, the radiation delivery is continuous over a period of time in days. The treatment time is greater than the repair half-time. Repair of sublethal damage (SLD) is possible up to 90%. The late-reacting normal tissue repair is greater than tumor and early reacting normal tissues. If LDR is combined with an external beam radiation therapy (EBRT), the overall treatment time is shorter for a given patient. There is a difference in the repair capacity of early- and late-reacting normal tissues (Figs 9.1A and B). The LDR maximizes the differential repair capacity of normal tissues. LDR also permits better apoptosis, i.e. organized cell death. The rationale for LDR brachytherapy is:

- Dose rate effect
- Reoxygenation
- Inverse dose rate effect.

In HDR brachytherapy, the treatment time is shorter compared to repair half-time of tissue. Only minimal repair occurs (~10%) and the cell kill in normal tissue is greater than that of tumors. Hence, HDR treatment has to be fractionated. If it is combined with an external radiation treatment, the overall treatment time is longer. However, there is an additional geometric sparing with optimization of dose, hence, it has therapeutic advantages over LDR brachytherapy. In addition, it offers complete radiation protection, with greater patient comfort.

However, the overall treatment time is longer than LDR brachytherapy and DNA repair takes place between fractions. Thus, the LDR advantage of lesser normal tissue damage

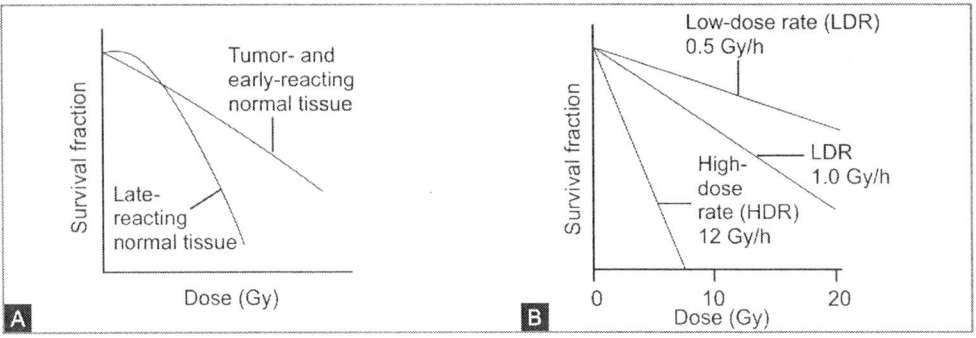

FIGURES 9.1A and B: Variation of survival fraction with dose. **A.** Differential repair capacity of normal tissue and tumor or early-reacting normal tissue; **B.** Survival curve for low-dose rate (LDR) and high-dose rate (HDR) brachytherapy.

and shorter overall treatment time are not achieved in HDR brachytherapy. Only fractionated HDR brachytherapy offers solution to the above issues. Then, the fractionated HDR brachytherapy has to balance the potentially unfavorable radiobiological factors namely:
- HDR dose reduction
- Dose per fraction
- Number of fractions.

The HDR can be safely used to replace LDR with respect to both tumor control and late effects, provided enough fractions are delivered and the total doses are reduced appropriately (Orton CG, 1994).

DOSE RATE EFFECT IN BRACHYTHERAPY

The LDR may be considered to be an infinite number of small fractions, delivered in long duration. There is no shoulder in the LDR survival curve and it is shallower than a HDR, single acute exposure curve (refer Fig. 9.1B). If different cell lines of varying radiosensitivity are irradiated with LDR, the survival curves fanout and show greater variation of slope of the curves, due to variation of inherent radiosensitivity (Figs 9.2A and B). There is also a range of repair times of SLD, as some cell line repair fast and some repair slowly. In contrast, the survival curves in HDR become closer as it overcomes the above factors more effectively. However, still there is an inherent variation of radiosensitivity.

The LDR intracavitary brachytherapy was started as a temporary implant at a dose rate of 0.5 Gy/h for 3 days treatment duration with radium-226 source. Later, manual afterloader with Cs-137 was used at a dose rate of 1.40 Gy/h for about 30 hours. Nowadays, HDR brachytherapy delivers the same treatment in three to five fractions with HDR (> 12 Gy/h) to achieve the same biological effect. In carcinoma cervix, the dose-limiting critical sites are rectum and bladder, which always receives lower dose than tumor in the above treatments. All the same, the total dose is reduced suitably for each dose rate employed in order to keep the biological outcome same.

The biological effect varies with dose rate in the range of dose rate used in interstitial brachytherapy. The maximum dose that can be delivered without undue damage to the surrounding normal tissue depends on the volume of tissue irradiated and dose rate. The dose rate is a function of number of implants and their geometric distribution. To achieve a desired biological effect, the total dose should be varied, depending upon the dose rate of the implant. Interstitial LDR brachytherapy was performed with a dose of 60 Gy delivered in 7 days and dose rate of 0.35 Gy/h (Paterson and Ellis). If the dose rate is higher, the total dose should be reduced to keep the biological endpoints same. For example, the above dose schedule is equivalent to 46 Gy in 3 days with a dose rate of 0.64 Gy/h.

Thus, the total dose varies with dose rate in a complex way. The variation of total dose as a function of dose rate is larger for late-responding normal tissues than early responding tissues, due to lower α/β values. It is also found that both tumor control and late effects vary with dose rate for a given total dose (Fig. 9.3). Patients treated with HDR are ended with necrosis clinically (~40%), compared with lower dose rate (~20%), irrespective of total dose. That is, necrosis rate is higher for the higher dose rate group at all dose levels.

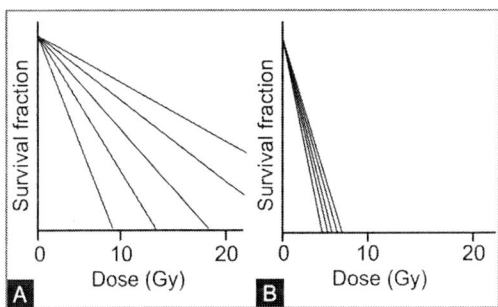

FIGURES 9.2A and B: Survival curve of cell lines of varying radiosensitivity. A. Low-dose rate (LDR); B. High-dose rate (HDR).

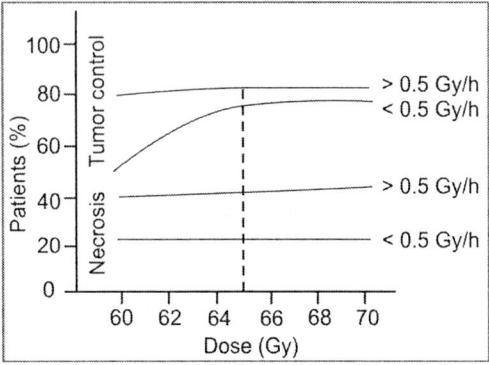

FIGURE 9.3: Effect of dose rate in necrosis and local control in a group of head and neck cancer patients with Ir-192 interstitial implants (Mazeron, 1990) [*Note:* The necrosis is higher at high-dose rate (HDR) for all range of doses. Tumor control also behaves in the same way for HDR. However, the tumor control for low-dose rate (LDR) behaves differently].

However, the low- and high-dose rate has differential effect in local tumor control. The percentage of patients with tumor control increased up to a total dose of 65 Gy for LDR and has little impact for above the dose range of 65–70 Gy. It is concluded that the local tumor control did not depend on dose rate, provided the total dose is sufficiently large. It is also found that for a given total dose, there are only few recurrences, if the radiation is delivered in multiple fractions with HDR. In HDR, Ir-192 is used whose half-life is 73.8 days, and the dose rate vary drastically, while in clinical use. Hence, such an isotope requires correction of total dose for a given dose rate.

In permanent implant, I-125 is commonly used with a dose of 160 Gy at the periphery of the implanted volume, with 80 Gy delivered in the first half-life of 60 days. The low energy (30 keV) photon of iodine has a relative biological effectiveness (RBE) of 1.5. It is equivalent to 120 Gy (80 × 1.5) of high-energy X-rays, delivered in 60 days. The tumor control depends on the cell cycle of the clonogenic cells. This type of implant is more suitable for slow-growing tumor such as carcinoma prostate.

Inverse Dose Rate Effect

As the dose rate decreases, there is an increase in cell killing, which is termed as inverse dose rate effect. This results from repair of SLD, redistribution and cell reproliferation. The survival curve is steeper with broad initial shoulder for an acute HDR exposure (Fig. 9.4). If the dose rate is reduced, repair of SLD take place and the cells are frozen in their cell cycle position, and do not progress in the cell cycle. Hence, the survival curve becomes shallower.

As the dose rate is reduced further, the curve become steeper in the inverse direction for a range of dose rate. Now, the cells progress through the cell cycle by redistribution, and accumulated in G2 phase, known as G2 block (Figs 9.5A and B). It is a radiosensitive phase and cells cannot divide. The corresponding dose rate is critical, which causes inverse dose rate effect.

If the dose rate is further lowered from the critical dose rate, the cells escape from the G2 block and divide, and proliferation of cells takes place during the radiation exposure, provided the exposure time is long compared with mitotic cycle, and dose rate is low. Thus, the mitosis offset the cell killing by the radiation and reduces the biological effect. Hence, the survival curve becomes shallow again in the reverse direction. Inverse dose rate effect do exists in some cell lines and it plays a key role in LDR brachytherapy.

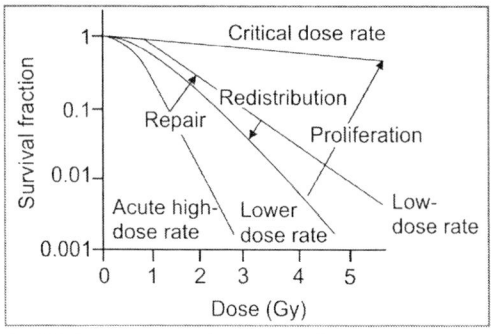

FIGURE 9.4: Inverse dose rate effect, increases cell killing at low-dose rate as seen in brachytherapy

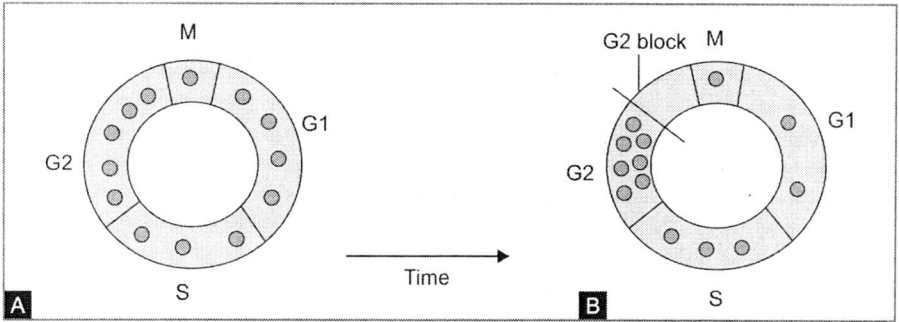

FIGURES 9.5A and B: Inverse dose rate effect in LDR brachytherapy. **A.** Asynchronous population of cells progress through the cell cycle; **B.** Cells get blocked in G2, become radiosensitive.

ROLE OF 4 Rs IN BRACHYTHERAPY

Repair

Repair of SLD occur due to the presence of DNA damage, detection of proteins and repair enzymes. Sufficient time is required to perform repair process. If the repair is not completed before another SLD, it may lead to lethal damage. Thus, there exists a difference in repair process in between low- and high-dose rates of radiation.

With low-LET radiation, the damage is minimal, for example, a single strand break (SSB) of DNA that can be readily repaired by enzymes. The repair process takes time of the order of minutes to several hours. Even for the low-LET radiation, some of the damage is irreparable and it may be due a double strand brake (DSB). Generally, cells undergo repair when they are exposed to low-LET radiation at LDR. With respect to high-LET radiation, no repair takes place, as it produces densely ionizing radiation causing severe damage to the cell.

At LDR, there is sufficient time for SLD repair, before the second damage occurs, and hence, the survival curve is shallower. Repair takes place both in normal and tumor tissue. The late-reacting normal tissues are more efficient in repair than tumor cells. The normal tissue may have both slow and fast or multiple components of repair. The repair half-time of normal tissue is of the order of 1.5–2 hours, whereas it is 0.5–1 hour for tumor. Though the repair half-time is lower in the case of tumor compared to that of normal tissue, it appears that the repair in tumor cells may not be in order due to the presence of mutation. Thus, tumor cells are preferentially killed compared with late-reacting normal tissue. Hence, even at LDR radiation, more tumor cells are killed than the normal cells.

The HDR radiation often causes damage before the first damage is repaired. For example, the first single strand brake caused by the HDR radiation is not likely to be repaired before the damage of second strand of DNA. Thus, HDR may not permit repair, unless the dose rate is reduced or dose is fractionated. Again, if the fractionation interval is of short duration, then the repair is prohibited during such HDR interval. However, if there is 12–24 hours interval between HDR fractions, SLD repair can occur. It may be incomplete in some tissues, where the repair is slow.

The repair rate varies during the course of treatment. It is believed that repair is fast with low doses, and become slower with high dose and dose rates. Thus, repair is the cause for dose rate effect in brachytherapy, and it is important in low dose and at LDR. It favors late-responding normal tissue than tumors.

Repopulation

Repopulation is of less importance in brachytherapy, which is usually a short duration treatment. Mostly, brachytherapy is combined with external beam radiotherapy, which is a longer treatment. Therefore, repopulation is mostly controlled by external beam radiotherapy rather than by brachytherapy. However, as the overall treatment time increases, the cells get repopulated. The cancer cells repopulate faster than late-reacting normal tissue.

In treatments involving LDR, the overall treatment time is shorter for a given patient, and it prevents repopulation during treatment and increases tumor control. In the case of HDR combined with EBRT, the overall treatment time is longer, which may permit repopulation and decrease tumor control. This is well proved in carcinoma cervix with LDR treatments. To handle HDR and tumor control effectively, all HDR-EBRT combined treatments should be completed within 56 days. However, repopulation plays a major role in permanent implant, which is a primary treatment and involves longer treatment time. In this type of treatments, it is better to match the half-life of the radionuclides to the proliferation rates of the tumor cells.

Redistribution

Redistribution has no role in brachytherapy, both in LDR and HDR. However, with sufficient fractionation interval, redistribution is possible in HDR. The inverse dose rate effect is observed in the in vitro experiments, but yet to be proved in the in vivo setup. The slower repair component in normal tissue is cell cycle dependent, which is capable of repairing more complex damages. This component is absent in many cancer cells; that is why, tumor is less efficient in repair.

Reoxygenation

The role of reoxygenation in brachytherapy is equivocal. The presence of long term and transient hypoxic cells are proved, at least in squamous cell carcinoma type of tumors. In LDR treatments, there is time for the transient hypoxia, which gets corrected during the treatment. During the interval of HDR fractions, the tumor shrinks and reoxygenation takes place. The capillary moves closer to the tumor and deliver more oxygen to the cells, permitting long-term hypoxia to get well oxygenated.

The oxygen-enhancement ratio (OER) is low (1.6–1.7) for LDR radiations, and 2–3 for HDR brachytherapy. At HDR, the cell survival curve is steeper and only little repair is possible. Oxygen is a dose modifier, which acts by interfering with repair process. Since the repair is less in HDR, the role of oxygen is also lesser. However, HDR treatments with hypoxic sensitizers may be beneficial. Thus, the presence or absence of oxygen has little impact on these cells in HDR brachytherapy. Probably, it may have some impact in permanent implant brachytherapy.

CLINICAL IMPLICATIONS

From the earlier discussion, one can recollect that there is a differential repair between tumor and late-reacting normal tissue that results a difference in their survival curves. It is also very clear that fractionation can split the curves and dose rate can vary the shape of the curves. The dose per fraction before and after the crossover points also has implications on the survival curves.

By referring Chapter 8 'Biological Basis of External Beam Radiotherapy', Figures 8.3, 8.4 and 8.5, it is well understood that the late-reacting tissue is above the tumor up to the crossover point, in the acute curve. After the crossover point, it gets reversed. If a dose per fraction below the crossover point is applied tumor get greater damage than the normal tissue. On the other hand, if the dose per fraction is applied above the crossover point, the curves get reversed and the normal tissue gets higher damage compared to the tumor. The same is hold good in brachytherapy also.

Consider the LDR survival curves illustrated in Figure 9.6A, for both acute and late reactions, relative to tumor. The separation between the tumor and normal tissue curves increases with decrease in dose per fraction or dose rate. At LDR, the survival curves of tumor and normal tissue are well separated, if the dose is applied before the crossover point of acute exposure. If the dose rate increases, the separation of curve will decrease. In this, late-reacting normal tissue curve is much shallow than the tumor. It demonstrates that the normal tissue survival is more than that of tumor cells. This is an ideal clinical situation for LDR treatment. The above discussion is valid for a dose rate of about 0.4 Gy/h.

If the LDR dose rate is doubled or too high, the normal tissue and tumor curves crossover point is shifted to higher doses (Fig. 9.6B). The separation of the curve also decreases. This is not a favorable clinical situation, since it causes more damage or equal damage to the late-reacting normal tissues.

The crossover point dose depends upon the LQ parameters used; to conclude that the use of LDR is advantageous to maximize cellular repair.

The HDR brachytherapy is usually performed with a number of fractions. If the HDR dose fraction is before the crossover point of acute exposure, the normal tissue and tumor curves are well separated (refer Fig. 8.4, Chapter 8 'Biological Basis of External Beam Radiotherapy'). In this, the normal tissue curve is above the tumor suggesting survival of normal tissue is more than that of the tumor or the cell-kill is more in tumor than that of normal cells. As the dose per fraction increases, the separation between the curve decreases. When the HDR fraction dose is applied above the crossover point of the acute exposure, the reversal trend of survival curve is observed (refer Fig. 8.5, Chapter 8 'Biological Basis of External Beam Radiotherapy'), i.e. the tumor curve is above the normal tissue, suggesting survival of tumor cells are more than that of normal tissue. Since it causes more harm to the late-reacting normal tissue, it is not a favoring clinical treatment.

However, there are exceptions where there is a geometric sparing of normal tissues. The geometrical sparing factor is defined as

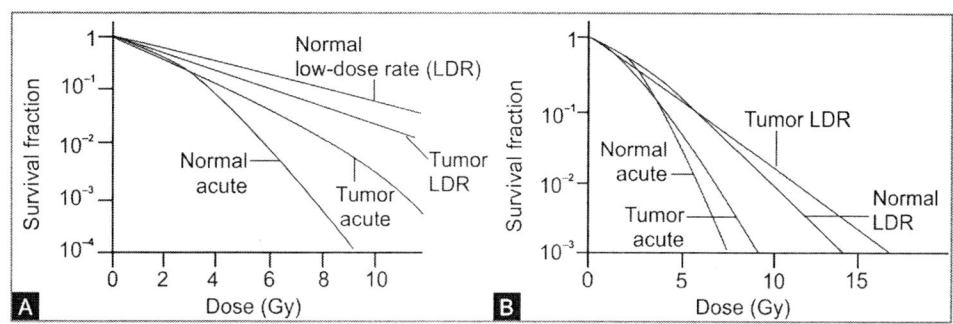

FIGURES 9.6A and B: Effect of LDR dose rate on survival fraction vs dose. **A.** LDR dose with a dose rate of 0.4 Gy/h falls before the crossover point of acute exposure, which provide better survival of late-reacting normal tissues (α/β for tumor is 10 Gy and normal tissue is 2.5 Gy); **B.** LDR with too higher dose rate (0.8 Gy/h), the tumor and late-reacting normal tissue crossover each other, predicting more damage to the normal tissue (α/β for tumor is 10 Gy and normal tissue is 2.5 Gy).

the ratio of effective normal tissue dose to the effective tumor dose. Normal tissue geometrical sparing offers the facility of shifting the crossover point to higher dose (Fig. 9.7). For example, if a normal tissue has geometrical sparing of 20%, the crossover point can be moved from 4 to 14 Gy. This permits the use of higher dose per fraction (6–9 Gy) in HDR brachytherapy. A typical example is carcinoma cervix in which the geometrical sparing of rectum is achieved with better retraction or backing. To conclude that the use of low dose per fraction is advantages to maximize cellular repair in HDR. However, higher dose per fraction can also be employed, since there is inherent geometrical sparing in HDR brachytherapy.

Conversion of LDR into HDR

The LDR treatment is a time tested one and required clinical experience is already available. The introduction of HDR treatment demands the equivalent LDR treatment schedule, in order to keep the biological affect same. Basically, the LDR survival curve is shallower than the HDR curve. The factors that influence the outcome in brachytherapy are:
- Dose
- Dose rate
- Fractionation.

The dose rate effect is much more important in LDR brachytherapy, whereas it is the fractionation in HDR. The dose rate in LDR brachytherapy permits the repair of sublethal damage in both tumor and normal tissue. Hence, repair component is the most suitable parameter to compare LDR and HDR.

If we assume that the repair of tumor and normal tissue is same, the dose of 60 Gy, in 72 hour is equivalent to 15–20 HDR fractions. If the repair of tumor cells is faster than normal tissue, the above HDR fractions reduced to 7 (Fig. 9.8). If there is geometrical sparing of normal tissue (G), the number of HDR fractions can be reduced still further, e.g. carcinoma cervix. To find the number of HDR fractions, the LQ model is used, which depends on repair rate constant (μ). Thus, G and μ decides the number of HDR fractions that are equivalent to LDR. It is found that the HDR fractions

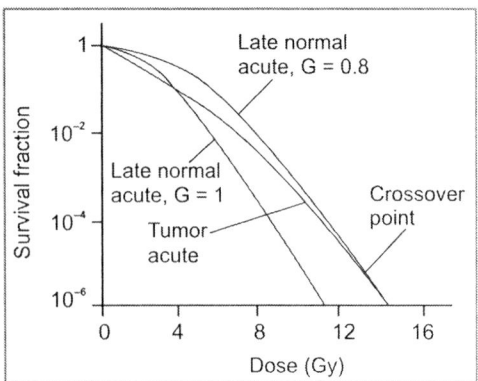

FIGURE 9.7: Geometrical sparing of late-reacting normal tissue pushes the crossover point to higher doses in an acute exposure. The crossover point moves from 4 to 14 Gy for a geometrical sparing of 20% for which G = 0.8.

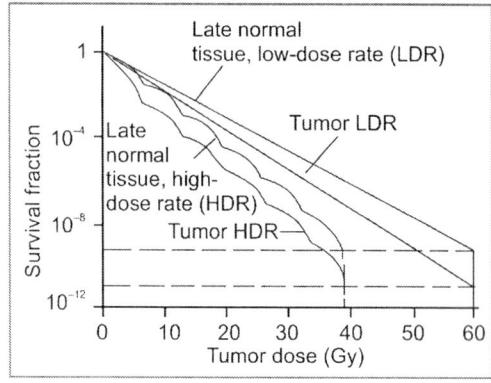

FIGURE 9.8: The LDR schedule of 60 Gy in 72 hours with a dose rate of 0.8 Gy/h, is equivalent to 6 HDR fractions of each 6.5 Gy per fraction. Both tumor and late-reacting normal tissue reactions are similar both in LDR and HDR, assuming a geometrical sparing of 0.8.

are equal to LDR only at a geometrical sparing of 0.8. HDR is better than LDR if the geometrical sparing is less than 0.8.

The geometrical sparing of normal tissue is common to both LDR and HDR. The HDR provides extrageometrical sparing, in addition to normal tissue sparing. Hence, the 60 Gy in 72 hours LDR schedule can be made equal to 5 HDR fractions. This will provide same local complication rate without reducing the local control. If the dose rate and dose per fraction is too high, normal tissue cell kill will be more than that of tumor. It is difficult to achieve local control without exceeding the normal tissue tolerance. Hence, the HDR schedule should have optimal dose per fraction and number of fractions. LQ model can be used to compare the LDR with HDR to get such optimal HDR treatment schedule.

Differential Repair Between Early and Late Normal Tissues

The survival curves of early and late responding normal tissues are not the same for fractionated brachytherapy. The survival curves of late responding normal tissue is curvier than that of early responding tissues (Figs 9.9A and B). The value of α/β is large for early responding tissue and effect of fractionation is less. In the case of late responding normal tissue, the α/β is small and the effect of fractionation is much significant. Hence, an increase in the number of fractions or a decrease in dose rate causes relatively bigger changes in late normal tissue than that of early-reacting tissue.

In other words, the increase of dose rate will increase the late effects much more than it will increase tumor control for a given dose. Similarly, decrease of dose rate will decrease the late effects much more than it will decrease the tumor control. Thus, therapeutic ratio increases as the dose rate decreases. This provides the rationale for LDR brachytherapy as follows:

1. Sublethal damage repair takes place during treatment, which maximize the differential response between early- and late-responding tissues.
2. The LDR treatment duration is lesser than EBRT, minimizing the effects of tumor repopulation, and hence, it resembles an accelerated treatment.

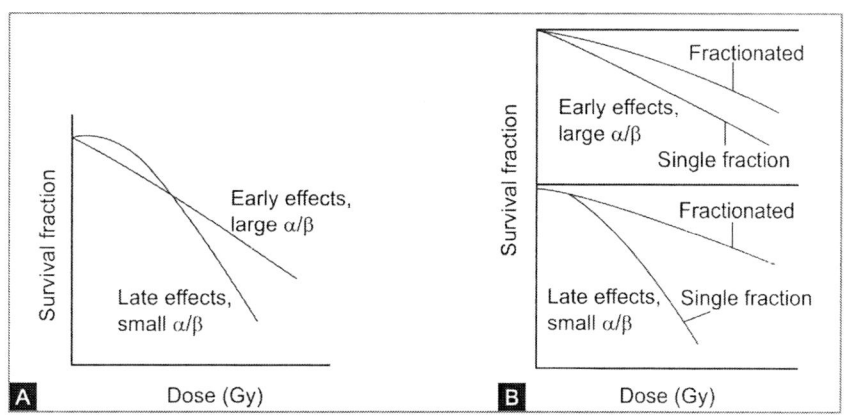

FIGURES 9.9A and B: Differential repair of early and late normal tissues. **A.** Variation of survival curve shape for early- and late-reacting normal tissue, the late normal tissue curve is curvier than the early reacting normal tissue, due to small α/β value; **B.** The effects of fractionation on late- and early-reacting tissue for a given dose; large change in effects in late normal tissue than in early reacting tissues.

3. Lower dose rate increases the therapeutic ratio, tumor cell repopulation if any, is only the limitation.

THERAPEUTIC RATIO IN BRACHYTHERAPY

The therapeutic ratio is related to dose rate in brachytherapy. It is constant at the lower and higher dose rate, and varies with dose rate in the middle (Fig. 9.10). The therapeutic ratio is almost constant over a dose rate range of 0.25–0.8 Gy/h, which belongs to the LDR region. That is why, LDR brachytherapy in earlier days has produced excellent clinical results. In the middle, the therapeutic ratio decreases with increase of dose rate. In this range, the biological effectiveness of the radiation varies with dose rate. As a result, the biological effectiveness of the given amount of dose varies with position of the source in the treated volume. The LDR remote after loaders may deliver such dose rate that falls in the middle range, and hence, caution is required to keep the normal tissue tolerance within control. Above 10 Gy/h, the therapeutic ratios vary little with the dose rate.

The therapeutic ratio again becomes constant over 100–1,000 Gy/h. Most of the HDR machine delivers a dose rate of 100–500 Gy/h at 1 cm and suits well in the above range. However, if the dwelling source is > 1 cm from the delivery distance, the dose rate falls and in turn changes the therapeutic ratio. If the therapeutic ratio at a point changes during the treatment, it is difficult to access the efficacy of that treatment. The HDR within the given volume gives therapeutic ratio in the flat region of the curve, is more advantageous clinically. This will simulate a condition that therapeutic effectiveness will follow the physical dose. To conclude, the therapeutic ratio decreases with increase of dose rate. However, fractionated HDR brachytherapy will improve the therapeutic ratio.

The flatness of HDR region suggests that it is independent of dose rate. The reason for this can be explained as follows. There are two damages, namely 'α' damage and 'β' damage. The 'α' damage is single track damage, causing DSB at the same location, which is independent of dose rate. The 'β' damage is caused by two independent events. The DSB is possible in different location or in the same location of the DNA strand. If enough time is available, the first side sublethal damage is repaired before the second side break at the same location. Hence, the HDR treatment should be of short duration, so that before the completion of the first strand break repair, the second strand break should happen at the same location. Thus, HDR executes all types of damages, irrespective of dose rate, before significant repair take place. This short duration depends upon the half-time for repair of sublethal damage. Hence, it is advantageous to complete the HDR treatments within half an hour.

The HDR brachytherapy is delivered in multiple fractions to handle therapeutic ratio. A HDR treatment schedule involving 4 fractions may improve the therapeutic ratio by 4.7%, compared to 3 fractions. If it is delivered in 5 fractions, it may yield another 4% over 4.7%. Thus, addition of fraction always

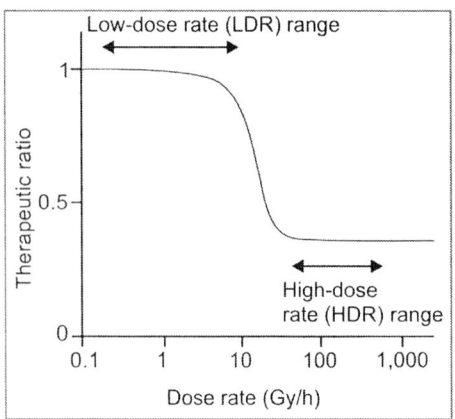

FIGURE 9.10: Relation between dose rate and therapeutic ratio, normalized to a dose rate of 0.1 Gy/h (α/β = 10 for tumor and 3 for late-reacting normal tissue).

improves therapeutic ratio further in HDR brachytherapy. If the HDR fraction is too higher, it is not only inconvenient to patient but also increase the cost of treatment. Hence, compromise need to be made between expected therapeutic ratio and clinical suitability. However, these fractions are entirely different from EBRT fractions, where the total numbers of fractions are higher, addition or deletion of one or two fractions may not affect the therapeutic ratio much.

The pulsed dose rate brachytherapy aims to maintain the repair advantage of LDR, while retaining the features of HDR such as radiation safety and optimization tool with a stepping source. A stepping source of activity 15–37 MBq gives a dose rate of 3 Gy/h, by pulsing for 10 minutes in an hour and 24 pulses per day. Though it is similar to LDR, but the dose rate is higher, which increases the risk of normal tissue than tumor. Hence, there may be some loss of therapeutic ratio in this technique.

PRACTICAL ISSUES IN RADIOTHERAPY

Volume Effects

Target volume is a critical factor in radiotherapy along with total dose and fraction scheme. Higher doses can be delivered, if the volume is too small, for a given fractionation schedule. Normal cells supposed to carry out certain functions, which are affected by radiation. Mostly normal cell could not regenerate from a single cell and its recovery is assisted by migration of unirradiated cells from the neighboring site, if the volume is small. Irradiation of large volume may have implications so that:
1. They have lesser opportunity to draw their functional reserve.
2. Larger irradiated volume makes its critical volume to exceed the upper dose limit.

However, these factors vary with tissue structure and the type of treatment. In general, normal tissue complication probability increases with dose and with irradiated volume. One should be aware of change in irradiated volume and its influence with tolerance dose. It is generally agreed that the tolerance dose is one, which may give not more than 5% complications. Of course, tolerance may also be affected by radiation sensitivity and fraction size. The severity of normal tissue depends on the volume of tissue that is irradiated, known as volume effect. However, it varies with organs and depends upon its:
- Structural organization
- Migration characteristics of surviving stem cells.

On the above basis, tissues can be divided into:
- Serial organization
- Parallel organization
- Serial-parallel organization
- Combination of the parallel and serial organization.

Serial organization is the one in which only a part of function of the organ volume is sufficient to carry out life cycle. Such organs have high reserve capacity and show resistance to partial radiation, even though they are radiosensitive, e.g. lung, kidney, etc. On the other hand, parallel organs have tube-like structure, and hence, even a partial damage can cause loss of function, e.g. spinal cord.

There are tissues, which have specific elements to carry out specialized function, e.g. brain. Radiation damage of even a small area will cause permanent damage to tissue function. The unaffected part may not carry out its function. In such tissue, the volume has only smaller impact on the tolerance dose. Irradiation of full volume may bring much more functional failure.

From the animal studies it is found that decrease in the field size require higher dose, for a given biological effect. Dose-response curve is found to be sigmoid. This is true in clinical treatment setting also. Changing the field size will have volume effect. The dose-response curve is steeper for series type tissue than parallel type tissue. The volume effect is less in the former and it is higher in the later.

Overall Treatment Time

Radiation treatment is performed over a period of 5–7 weeks, which is sufficient for cell proliferation, both in tumor and normal tissues. Early-reacting normal tissue have high proliferation rate; they have ability to withstand radiation due to extensive repopulation. If the overall treatment time (OTT) is shorter, repopulation may be less and severe reaction will occur. On the other hand, late-reacting tissues have slow proliferation. The OTT has lesser impact on these tissues. Usually, for a slight change in OTT, there is no need to adjust the radiation dose.

The tumor repopulation varies widely. Most of the human tumor grows very slowly. When a tumor is damaged and begins to shrink, there is an increase in repopulation rate. The doubling time for repopulation is found to be less than 7 days. Hence, OTT has very much impact on tumor response. Even a extension of 1 week will influence the probability of tumor control. On the other hand, even a slight shortening of OTT will improve the chances of tumor control. Both for tumor and early-reacting normal tissues, there is a period for the commencement of repopulation, which is usually 2–3 weeks.

Effect of Treatment Delay

The time taken to begin radiotherapy treatment after diagnosis is known as treatment delay or waiting period. The delay in treatment has direct influence on local control probability of tumor. This can be estimated based on the increase in the number of colonies during the waiting period. If N_0 is the initial number of colonies, then the number of colonies (N) at a given time (T) is:

$$N = N_0 2^{T/T_{eff}} \quad \text{-----(1)}$$

Here, T_{eff} is the pretreatment tumor doubling time that predicts the growth rate. Tumor growth during the waiting period can be modeled based on the effect of doubling time of tumor colonies. Though the treatment delay effect is predictable numerically, it is very difficult to observe in clinical practice. However, the delay between surgery and radiotherapy is very critical, especially in head and neck cancer. The effects of treatment delay on various organs are detailed below.

Head and Neck Cancer

In head and neck cancer, treated with radiotherapy alone, the tumor control probability (TCP) decreases with waiting period (Fig. 9.11). The effect is seen more on larger tumor (T3, T4) than small size tumors (T1). Any delay up to 3 months, have smaller, but non-negligible effect on treatment outcome. The reduction of TCP is found to be 0.8–1.4% per month. In the case of large size tumors, it is 1.7–3.0% per month delay. In head and neck treatment involving surgery and radiotherapy, the waiting time between surgery and radiotherapy is very critical. The TCP decreases much steeper and the reduction

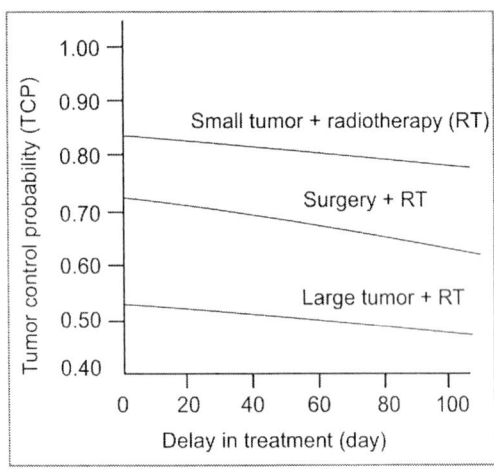

FIGURE 9.11: Effect of treatment delay on tumor control probability is based on a regime of 70 Gy/35# in 7 weeks.

is about 3–6% per month delay. This is because of the faster tumor-doubling time after surgery. Treatment regime with smaller total dose (e.g. CHART regime) have greater decrease in TCP over the waiting period. This is true for both radiotherapy and surgery plus radiotherapy patients.

Breast Cancer

Breast cancer patients, with postsurgery are found to decrease the local control by 0.8–2.9% per month delay. This depends upon the treatment regime used. Regime with lowest total dose may have greater reduction. The effect is also greater, about 2–7% per month delay for fast-growing tumor. The fast-growing tumor, with lower total dose or relatively longer treatment time has much reduction in TCP. About 21% of the breast cancer has the doubling time < 2 weeks, hence it is important to keep the duration between surgery and radiotherapy as short as possible.

Cervix Cancer

The pretreatment doubling time of cervical cancer is generally slow and it is about 119 days. It is difficult to calculate the TCP for cervix cancer, since the dose is delivered to a fixed point (point A), which is not a representative of tumor. The dose to the tumor is non-uniform, due to steep dose gradient, especially with brachytherapy. Cervical cancer patient often is given chemotherapy along with radiotherapy. The local control reduction is reported as 0.6–0.9% per month delay. A treatment delay of up to 2–3 months is not expected to cause significant effect on TCP.

Reirradiation

Radiation is capable of causing second malignancy. The dose limit, which can cause second malignancy, is uncertain. Multiple beams or IMRT may increase the irradiated volume of normal tissue and increase the risk of second malignancy. It is found that the chance of second malignancy is 1 in 10 of irradiated patients. The causes are:
1. Lifestyle factors, which caused the first malignancy may continue and be responsible.
2. Genetic predisposition.
3. Radiotherapy or chemotherapy.

The malignancy caused by radiotherapy may be within the first field, or close to the field or at remote locations. Since children are more sensitive to radiation than adults, the chances of second malignancy is higher in children.

If the tissue is already irradiated with higher dose, then surgery option is remote. If the radiation tolerance is exceeded in the first radiation, then second radiation cannot be executed because the tissue has already lost its function. On the other hand, if the first treatment is within or less than the tolerance of the organ, then second irradiation can be planned. While planning such irradiation, one should consider the following factors:
- Dose and volume of the first irradiation
- Overlaps of second treatment field with the first treatment field
- Gap between the two irradiation
- Other modalities of treatment such as brachytherapy or chemotherapy during the first treatment
- Nature of critical organs involved.

Based on the above factors, conformal radiotherapy such as intensity modulated radiotherapy, stereotactic body radiotherapy, brachytherapy or proton therapy may be planned. Early-reacting normal tissues are highly proliferating, can tolerate well reirradiation and capable of restoring damage, if any. On the other hand, late-reacting normal tissues are slowly proliferating and unable to tolerate the damage.

Animal data suggests that retreatment with reduced dose is possible both in lung and spinal cord. However, there is some difficulty in transferring this data to humans.

The human data on reirradiation is very small; dose and time interval between irradiation vary widely. Reirradiation may be recommended with reduced doses, but the morbidity may be higher. Data is available with 50–60 Gy dose of reirradiation in head and neck patients, but lower doses are found ineffective. A local control of about 30% is achievable with such doses; however, series complications such as subcutaneous fibrosis, trismus and impaired hearing may occur. Reirradiation of recurrent nasopharyngeal cancer with 35% local control for T1 stage and 11% for T3 stage was reported. The 5-year local control for T1 and T2 recurrences was found in patients treated with > 70 Gy_{10} biologically effective doses (BEDs).

Reirradiation of cervical cancer was not encouraging and the local control is very poor. However, it can be improved up to 60% of long-term survival, in small tumor volume, second primary malignancy and with brachytherapy treatment. Results with large tumor volume, recurrent cancer and external beam treatment are not encouraging. Retreatment of recurrent lung cancer with palliative dose of 30 Gy in 3 weeks showed improvements in symptoms. However, the retreatment doses must be lesser than the first treatment dose. Recurrent metastatic brain lesions, rectal cancer and breast cancer have also showed improvement in symptoms with palliative treatment. To conclude, reirradiation is possible in some sites as effective palliation without high morbidity. If full dose is delivered, good local control can be achieved with increased level of morbidity.

BIBLIOGRAPHY

1. Colin G Orton. Fractionation: Radiobiologic Principles and Clinical Practice. In: Faiz M Khan (Ed). Treatment Planning in Radiation Oncology, 2nd edition. Philadelphia, USA: Lippincott Williams & Wilkins; 2007.
2. Colin G Orton. Radiobiology. In: Subir Nag MD (Ed). High Dose Rate Brachytherapy. Armonk, NY: Futura Publications Company Inc; 1994.
3. Colin G Orton. Radiobiology. In: Subir Nag MD (Ed). Principles and Practice of Brachytherapy. Armonk, NY: Futura Publications Company Inc; 1997.
4. Eric J Hall, Amato J Giaccia. Radiobiology for the Radiologist, 7th edition. Philadelphia, USA: Wolters Kluwer/Lippincott Williams & Wilkins; 2012.
5. Loredana G Marcu. Eva Benzak. Recent Advances and Research Updates: Medical Physics and Radiobiology. Trivandrum, India: Researchman Publishers Pvt Ltd; 2011.
6. Mayles P, Nahum A, Rosenwald JC. Handbook of Radiotherapy Physics-Theory and Practice. New York: Taylor & Francis; 2007.
7. Mould RF. Brachytherapy from Radium to Optimization. Netherlands: Nucletron BV; 1994.
8. Phillip M Devlin. Brachytherapy Applications and Techniques. Philadelphia, USA: Lippincott Williams & Wilkins; 2007.
9. Sanjay S Supe. An Overview of Time Dose Fractionation Models; NSD to BED. Pol J Med Phy Eng. 2006;12(4):165-201.

Time, Dose and Fractionation Models

CHAPTER 10

ISOEFFECT MODELS

Radiotherapy is more effective, if it is optimally fractionated and delivered, over a period of time. Various fractionation schemes were established in different clinical centers with considerable difference in treatment and doses. These differences in treatment schedule for the same condition of diseases led to the search for the rationalization of fractionation scheme for their biological effectiveness. It was felt that some basis of comparison on a common scale, which will unify the variable techniques, would be of great value. Since normal tissue tolerance is the limiting factor in radiotherapy, it was decided to compare the measure of radiation damage in the normal tissue.

Several workers attempted towards this and evolved several concepts, which were useful to the new entrants in the field of radiotherapy. The above such concepts are known as time, dose and fractionation models. These concepts do not explain all the complex biological phenomena that take place at cellular level during and after the course of radiotherapy. But, it provides a simple and convenient way of evaluating and comparing the biological effectiveness of various treatment schedules. The time, dose and fractionation models have been in clinical practice for several decades. It is mainly used to:

1. Calculate the total dose required to maintain the same biological effectiveness, when a new fractionation scheme is employed.
2. Compare the efficacy of treatment techniques that differ in dose per fraction, number of fractions and over all treatment time.
3. Design new fractionation schemes.
4. Development of bio-effect dosimetry with combined modalities.
5. Retrospective analysis of clinical data.
6. Derivation of therapeutic ratios and optimization.

Over the periods, several time, dose and fractionation models are developed, namely:
- Cube root rule
- Nominal standard dose (NSD)
- Cumulative radiation effect
- Linear-quadratic model.

CUBE ROOT RULE

In 1944, Strandquist had studied the skin reactions including skin erythema and skin tolerance, and tried to correlate the dose to overall treatment time (OTT). He plotted the skin reactions in a log-log graph, between total dose and over all treatment time. These curves are shown to be straight lines called isoeffect curves or Strandquist lines (Fig. 10.1). Strandquist proposed that the dose (D) needed to reach the normal tissue tolerance is proportional to the cube root of OTT (T):

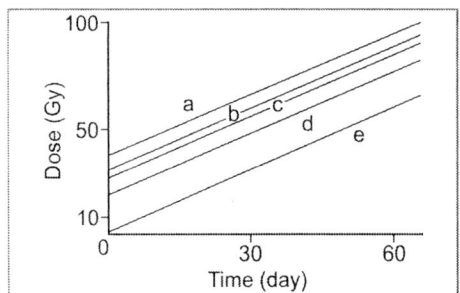

FIGURE 10.1: Isoeffect curves in a log-log plot between total dose and treatment time (**a**, necrosis; **b**, skin carcinoma cure; **c**, moist desquamation of skin; **d**, dry desquamation of skin; **e**, skin erythema).

$$D_n \propto \sqrt[3]{T} = K_1 T^{0.33} \quad \text{-----(1)}$$

where,

D_n is the dose to normal tissue, and K_1 is a constant. These curves are drawn based on the treatment data consisting five fractions per week. Hence, T includes the number of fractions and the OTT is a linear function of number of fractions (N). Hence, the dose is also proportional to cube root of N. However, it fails to explain what happened if other than five fractions per week was delivered.

Cohen (1968) studied the above straight line curves and suggested that they have a slope of 0.22, for the tumor cure dose (squamous cell carcinoma) and 0.33 for various degree of skin reactions such as necrosis, moist desquamation, dry desquamation and erythema. Hence, the relation between dose and T for tumor is given by:

$$D_t = K_2 T^{0.22} \quad \text{-----(2)}$$

where,

D_t is the dose to tumor and K_2 is a constant. The exponent of T in the above two equations represent the average values of repair and recovery capacities of normal and tumor tissues.

NOMINAL STANDARD DOSE

Frank Ellis (1969) realized that the cube root relation was a result of repair and repopulation. According to him, the repair for normal cells is larger than that of tumor and it is due to lack of homeostatic control in tumor cells. He also believed that homeostatic recovery was a long-term recovery, whereas the intracellular recovery was a short-term recovery. The intracellular recovery has a half-life of 1–2 hours, could be completed after few hours of irradiation. Studies have shown that the isoeffect dose is not only a function of time but also a function of number of fractions. It was demonstrated that dose was predominantly dependent on the number of fractions. Thus, the relationship between recovery and fractionation was established.

According to Ellis, the exponent of time for normal cell in equation (1) has two parts, namely:

- Elkind type intracellular recovery
- Homeostatic recovery.

He believed that recovery is functions of both number of fractions and over all treatment time. It is also believed that fractionation was twice as important as OTT in influencing dose at which skin reaction occurs. Therefore, he divided the Strandquist slope 33 into 11 and 22, where 11 is the exponent of OTT that represents the homeostatic recovery. The slope 22 is the exponent of number of fractions, rather than treatment time that represents the intracellular recovery.

$$D \alpha\ T^{0.11} T^{0.22}$$

He postulated the isoeffect dose as a function of OTT (T) and number of fractions (N).

$$D \alpha\ T^{0.11} N^P$$

He wanted to replace $T^{0.22}$ by N^P, which would give the same amount of recovery as given by factor, $T^{0.22}$ based on Strandquist observations. Strandquist observations were related to five fractions per week and the standard treatment was to give 30 fractions of 2 Gy each in 40 days. He obtained the value of P as 0.24 from the relation $30^P = 40^{0.22}$.

$$D = NSD\ T^{0.11} N^{0.24}$$

where,

NSD is a constant of proportionality for a specific level of skin damage. Thus, exponent of N is not 0.22, because T is not exactly equal to N, but (7/5) N, for patients treated for five days per week. It depends on which day of the week the treatment is started. Hence, N is assigned an exponent of 0.24, which give the best fit to the Strandquist curve for patient treated for 5 days per week. Truly speaking, the NSD refers a level of biological effect as given below:

$$NSD = D\, T^{-0.11}\, N^{-0.24} \quad\text{-----(3)}$$

The NSD can be used to compare two treatment regimens, by comparing the NSD values. However, NSD cannot be added similar to physical doses, since it is not a linear function of N. The total dose D = N × d, where N is the number of fractions and d is dose per fraction and this can be substituted in the above equation as:

$$NSD = N \times d\, (T^{-0.11}\, N^{-0.24})$$
$$= d\, (N \times N^{-0.24} \times T^{-0.11})$$
$$= d\, (N^{0.76} \times T^{-0.11})$$
$$= d\, (N^{0.65} \times N^{0.11} \times T^{-0.11})$$
$$NSD = d \times N^{0.65} \times (T/N)^{-0.11} \quad\text{-----(4)}$$

where,

(T/N) is a constant and its unit is rad equivalent therapy (ret). Ellis assumed that the cancer cells are not susceptible to homeostatic control. The coefficient 0.24 represents the cell damage and internal recovery, while the regression coefficient 0.11 represent for homeostatic influence of normal tissues only. Since it is based on skin reaction data, it is not addressing late effects of radiation. The value of NSD differs center to center, due to:
1. The dosimetry may not be identical in all centers.
2. Variation in the complication rate.
3. Variation of patient material such as age, sex, race, etc.

The NSD formula is applicable only for those treatment schedules, which lead to full tolerance of normal tissues. The effectiveness of treatment, which do not result in the delivery of tolerance dose to normal connective tissue that can be evaluated in terms of partial tolerance (PT), defined by Ellis (1969) as:

$$PT = NSD \times \frac{n}{N} \quad\text{-----(5)}$$

where,

N is the number of fractions, which results in full tolerance and n is the number of such dose fractions actually given. The partial tolerance is additive in nature, unlike NSD.

Limitations of NSD

The formula does not take into account all complex processes that take place during or after the course of radiation. Hence, the formula cannot be taken as a fundamental law of nature; it is merely intended to provide a simple and convenient method, of relating dose, number of fractions and OTT. It is applicable only for those treatment schedules, which lead to full tolerance reactions of the normal tissue. The formula hold good only for a treatment time T within the range of 3-100 days and for 4 or more fractions and the fraction should be spaced by at least 16 hours. Further, the fractionation should be spaced evenly over the treatment time. If there is a change in fractionation or if there is a gap in the treatment, it should be considered as separate entity. The last, but not the least is that the NSD concept does not take into account the effect of volume of normal tissue irradiated.

CUMULATIVE RADIATION EFFECT

The cumulative radiation effect (CRE) was suggested by Kirk et al (1971). It is a generalized form of the empirical function of NSD. The CRE formula is identical and numerically comparable with the NSD formula at the limit of normal connective tissue tolerance. More over CRE formula is defined at the subtolerance levels of radiation damage on normal tissue. They proposed CRE for external beam therapy, which can be obtained from equation 4 as:

$$CRE_F = X^{-0.11} \times d \times N^{0.65} \quad \text{-----(6)}$$

where,

$X = T/N$, the average intertreatment interval. T is the OTT, N is the number of fractions, d is the dose per fraction and the unit of CRE is reu (radiation effective unit).

They also deduced CRE for brachytherapy treatment both for short- and long-lived radioisotopes. In brachytherapy, it is stated that the generation of lethal and sublethal damage (SLD) is a continuous process and the repair and recovery takes place simultaneously during irradiation. The treatment time in many cases is in days, which is of the order of tens of repair half-times. Hence, the relation between dose (D_c) and treatment time for brachytherapy is given by:

$$D_c = K_3 T^V$$

where,

The exponent $V = 0.29$ and K_3 is a normalizing constant used to equate the CRE at full tolerance of tissues in external beam therapy to that of brachytherapy. Based on this, Kirk et al established the equation for brachytherapy as:

$$CRE_C = K_3 D T^{-0.29}$$
$$CRE_C = K_3 R \times T \times T^{-0.29}$$
$$CRE_C = K_3 R T^{0.71} \quad \text{-----(7)}$$

where,

R is the dose rate in cGy/day, T is the treatment time in days and $R \times T = D$. The value K_3 is stated as 0.54 and later modified as 0.77. They also suggested a volume correction factor, if the volume of brachytherapy and that of external beam therapy is different. Its limitation includes non-additive nature of CRE values and volume correction factor.

TIME-DOSE-FRACTIONATION MODEL

Orton and Ellis (1973) developed the time, dose and fractionation concept to make use of partial tolerance of normal tissue more practicable. The NSD empirical model can be made additive by rising power $1/0.65 = 1.54$, on both sides of the above equation (4). Thus, the NSD equation is modified and written as time, dose, fractionation (TDF) as:

$$TDF \propto NSD^{1.54}$$
$$= 10^{-3} \times NSD^{1.54}$$

Substituting equation 4:

$$TDF = N.d^{1.54} (T/N)^{-0.17} \times 10^{-3}$$

where,

10^{-3} is a scaling factor to make TDF 100, which corresponds to maximum skin tolerance. For example, a TDF value of 99 is approximately equal to 60 Gy in 30 fractions over a period of 6 weeks. The equation can be rewritten by incorporating SI units, i.e. dose in Gy and T in days:

$$TDF = 1.19 \times N. d^{1.54} \times (T/N)^{-0.17} \quad \text{-----(8)}$$

The bioeffective doses in TDF units are linearly additive, since TDF is a linear function of N. Whenever there is an interruption in radiotherapy treatment, repopulation may occur during the rest period. Ellis proposed a decay factor for giving allowance for such repopulation as:

$$\text{Decay factor} = \left(\frac{T}{T+R}\right)^{0.11}$$

where,

T is the number of days from the starting of the course of radiotherapy to interruption and R is the number of days of rest. The overall TDF must be multiplied with the decay factor, to obtain the true TDF value, which compensate for the rest. The greatest advantage of TDF is that it is independent of any specific NSD value and readymade table is available for easy reference.

The authors have also applied the TDF concept to brachytherapy. They presented TDF table for various values of dose rate and time of application. Since TDF is additive, evaluation of TDF is very easy. It is also useful in combined treatment involving external beam therapy and brachytherapy.

Worked example 1

Calculate the TDF value for standard external beam treatment of 60 Gy, 2 Gy per fraction delivered in 30 fractions.

Here,
N = 30 fractions
d = 2 Gy
T = 42 days

$$TDF = 1.19 \times N \times d^{1.54} \times \left(\frac{T}{N}\right)^{-0.17}$$

$$TDF = 1.19 \times 30 \times 2^{1.54} \times \left(\frac{42}{30}\right)^{-0.17}$$

$$= 35.7 \times 2.91 \times 0.94 = 97.65$$

Worked example 2

A patient is planned with a standard external beam treatment of 60 Gy, 2 Gy per fraction delivered in 30 fractions. He developed acute reactions after 25 fractions and is given 2 weeks rest. Calculate the dose per fraction to complete the treatments in additional five fractions:

$$TDF = 1.19 \times 25 \times 2^{1.54} \times \left(\frac{35}{25}\right)^{-0.17}$$

$$= 29.75 \times 2.91 \times 0.94 = 81.37$$

The decay factor to compensate rest period =

$$\left(\frac{T}{T+R}\right)^{0.11} = \left(\frac{35}{35+14}\right)^{0.11} = 0.964$$

The TDF after the rest period = 81.37 × 0.964 = 78.45

The planned TDF is

$$= 1.19 \times 30 \times 2^{1.54} \times \left(\frac{42}{30}\right)^{-0.17}$$

$$= 35.7 \times 2.91 \times 0.94 = 97.65$$

The remaining TDF = 97.65 − 78.45 = 19.20

The dose per fraction (d) to complete the treatment in 5 fractions is:

$$19.20 = 1.19 \times 5 \times d^{1.54} \times \left(\frac{7}{5}\right)^{-0.17}$$

$$= 5.95 \times d^{1.54} \times 0.94 \text{ or}$$

$$d = 2.23 \text{ Gy}$$

LINEAR QUADRATIC MODEL

Linear quadratic model is derived from the linear quadratic cell survival theory, based upon the fundamental mechanism of interaction of radiation with biological systems (Chapter 4). As per the linear quadratic theory, the survival fraction after a single fraction is:

$$-\ln S = (\alpha d + \beta d^2) \quad \text{-----(9)}$$

If N fractions are delivered, the survival fraction after exposure:

$$-\ln S = N(\alpha d + \beta d^2) \quad \text{-----(10)}$$

where,

d is the dose per fraction in Gy. The above equation accounts only the repair mechanism. However, there is repopulation of cells by the early responding normal tissues and tumors. The tumor repopulation is an exponential function of time and it is given by the equation $N = N_o e^{\lambda T}$, where N is the number of clonogens at a time t and N_o is the initial number of clonogens. The λ is a constant related to the potential doubling time of tumor (T_{pot}):

$$\lambda = \frac{0.693}{T_{pot}} \quad \text{-----(11)}$$

Then, the survival fraction will be increased by $\dfrac{0.693}{T_{pot}} \times T$ and the final equation is given by:

$$-\ln S = N(\alpha d + \beta d^2) - \dfrac{0.693T}{T_{pot}} \quad \text{-----(12)}$$

Dividing the entire equation by $-\alpha$, the equation becomes

$$-\dfrac{\ln S}{\alpha} = N\left(d + \dfrac{d^2}{\alpha/\beta}\right) - \dfrac{0.693T}{\alpha T_{pot}} \quad \text{-----(13)}$$

The left hand side is called biologically effective dose (BED) or extrapolated response dose (ERD), which refers to the quantum of surviving cells per 'α' event, following irradiation. Thus, term BED predicts radiation effects and its local control quantitatively. The final equation is written as (Barendsen and Fowler):

$$\text{BED} = -\dfrac{\ln S}{\alpha} = N\left(d + \dfrac{d^2}{\alpha/\beta}\right) - \dfrac{0.693T}{\alpha T_{pot}} \quad \text{-----(14)}$$

The potential doubling time vary from patient to patient and difficult to determine its value for individual tumors. Hence, it is replaced by a generic repopulation parameter k, which can be determined by clinical data analysis in a group of patients of specific diseases:

$$\text{BED} = N\left(d + \dfrac{d^2}{\alpha/\beta}\right) - k(T - T_k)$$

$$= N \times d\left(1 + \dfrac{d}{\alpha/\beta}\right) - k(T - T_k) \quad \text{-----(15)}$$

where,

T_k is the repopulation kick-off time after a radiation treatment and T is the OTT in days. It is believed that tumor repopulation is negligible up to 21–28 days after the beginning of treatment. Thus, T_k refers the repopulation start up time and it is repopulated at the rate of k Gy per day, where k = 0, for $T < T_k$. The OTT is very important for rapidly growing tumors. In head and neck cancer, the local control decreases about 1.4% for each day of prolonged treatment. Similarly, it is about 0.5% per day for carcinoma cervix. The product N × d refers the total dose (D) and the term $\left(1 + \dfrac{d}{\alpha/\beta}\right)$ is called relative effectiveness (RE). Hence, the BED is the product of total dose and relative effectiveness (D × RE).

The LQ model is useful for the study of several interesting problems both in external beam radiotherapy and brachytherapy, especially in converting the high-dose rate (HDR) treatments into low-dose rate (LDR) equivalent. The BED used in the above discussion is based on the tissue characteristics, which depends on the α/β ratio. This ratio will vary with tissue types (Table 10.1). Fowler (1990) has suggested a α/β value of 10 and 3 for tumor and late reacting normal tissue respectively. Hence, the BED comparisons should be carried out for tumor, early reacting normal tissue and late reacting normal tissue separately. This is useful in comparing a standard fractionation scheme with a new scheme. For example, if D_{STD} is the total dose and d_{STD} is the dose per fraction in the standard dose fraction scheme, for a given tissue, then this can be compared to a new fractionation scheme with a total dose of D and dose per fraction, d as follows:

$$D\left(1 + \dfrac{d}{\alpha/\beta}\right) = D_{STD}\left(1 + \dfrac{d_{STD}}{\alpha/\beta}\right)$$

$$\dfrac{D}{D_{STD}} = \dfrac{d_{STD} + (\alpha/\beta)}{d + (\alpha/\beta)} \quad \text{-----(16)}$$

Table 10.1: The α/β ratio for early and late reacting normal tissues from multifraction experiments

Effects	α/β (Gy)
Early effect	
Skin	9–12
Jejunum	6–10
Colon	10–11
Testis	12–13
Callus	9–10
Late effect	
Spinal cord	1.7–4.9
Kidney	1.0–2.4
Lung	2.0–6.3
Bladder	3.1–7.0

Worked example 3

Calculate the BED value for a conventional external beam treatment of 60 Gy, 2 Gy per fraction delivered in 30 fractions (assume tumor α/β = 10 Gy, T_k = 28 days and k = 0.6 Gy/day, early reacting normal tissue α/β = 10 Gy, T_k = 0 days and k = 0.25 Gy/day, and late reacting normal tissue α/β = 3 Gy, T_k = 28 days and k = 0 Gy/day).

Here, N = 30, d = 2 Gy and T = 42 days, the LQ equation is:

$$BED = N \times d \left(1 + \frac{d}{\alpha/\beta}\right) - k(T - T_k)$$

1. **Tumor:**

$$BED = 30 \times 2 \left(1 + \frac{2}{10}\right) - 0.6(42 - 28)$$

$= 60 \times 1.2 - (0.6 \times 14) = 72 - 8.4 = 63.6$ Gy

2. **Early reacting normal tissue:**

$$BED = 30 \times 2 \left(1 + \frac{2}{10}\right) - 0.25(42 - 0)$$

$= 60 \times 1.2 - (0.25 \times 42) = 72 - 10.5 = 61.5$ Gy

3. **Late reacting normal tissue:**

$$BED = 30 \times 2 \left(1 + \frac{2}{3}\right) - 0(42 - 28)$$

$= 60 \times 1.66 = 100$ Gy

The BED value is always greater than the physical dose at any distance and its unit is Gy. Similarly, the BED value of late reacting normal tissue is higher than that of tumor. However, the late normal tissue BED value changes more over distance, compared to tumor BED, which will be discussed in the later paragraphs.

Worked example 4

An accelerated fractionation scheme consisting of 40 treatments, twice daily for 4 weeks is to be equivalent to a standard external beam treatment of 60 Gy, 2 Gy per fraction delivered in 30 fractions. What is the required dose per fraction for the treatment?

Here, N = 40, d = ? Gy and T = 28 days and the LQ equation is:

$$BED = N \times d \left(1 + \frac{d}{\alpha/\beta}\right) - k(T - T_k)$$

1. **Tumor:**

$$BED = 40 \times d \left(1 + \frac{d}{10}\right) - 0 = 4d^2 + 40d$$

Since there is no repopulation, this has to be equated to the BED of 63.6 Gy_{10} of 60 Gy in 30 fractions at 2 Gy per fraction (worked example 3), hence, dose per fraction (d) is obtained as follows:

$4d^2 + 40d - 63.6 = 0$

Solving the equation, we get

d = 1.40 Gy per fraction.

2. **Early reacting normal tissue:**

$$BED = 40d \left(1 + \frac{d}{10}\right) - 0.25(28 - 0)$$

$= 4d^2 + 40d - 7$

This has to be equated to 60 Gy in 30 fractions, 2 Gy per fraction, for which the BED = 61.5 for early normal tissue reactions, to find the dose per fraction (d),

$4d^2 + 40d - 7 = 61.5$ or
$4d^2 + 40d - 68.5 = 0$,

Solving the equation, we get
d = 1.49 Gy per fraction.

3. **Late reacting normal tissue:**

$$BED = 40 \times d \left(1 + \frac{d}{3}\right) - 0$$

Since no repopulation, correction is required. This has to be equated to 60 Gy in 30 fractions, 2 Gy per fraction, for which the BED = 100 for late normal tissue reactions, to find the dose per fraction (d),

$$40d + \frac{40d^2}{3} = 100$$

or $40d^2 + 120d - 300 = 0$

Solving the equation, we get
d = 1.6 Gy per fraction.

In conclusion, the dose per fraction required are 1.40, 1.49 and 1.6 Gy per fraction for tumor, early reacting normal tissue and late reacting normal tissue respectively. Since the late reacting normal tissue fraction is higher than that of tumor, any fraction size from 1.4 to 1.6 Gy is clinically acceptable.

LQ Model for Brachytherapy

Modeling studies have attempted to equate the biological effects of various brachytherapy techniques to quantify the potential gain in tumor control offered by the source dwell time optimization. It also verifies methods, which reduce over doses of the critical target structures. Based on the linear quadratic model, the BED for HDR treatments involving a dose d and fractions of N =1 is given by the relation:

$$BED_{HDR} = d\left(1 + \frac{d}{\frac{\alpha}{\beta}}\right) - \frac{0.693T}{\alpha T_{pot}}$$

----- (17)

The right hand side is the repopulation term, which is generally not used in HDR calculations. HDR is a short treatment time and no repair of potential 'β' damage is possible, and hence, the BED for 'n' number of HDR fractions is given by:

$$BED_{HDR} = n \times d\left(1 + \frac{d}{\frac{\alpha}{\beta}}\right)$$

----- (18)

In the case of LDR, the dose rate is low and the duration of treatment is long, of the order of several hours. This has two consequences, namely:

1. A fraction (G) of 'β' damage begins to decrease.
2. Repair occurs during the treatment.

In addition, repopulation also takes place, which is important for the tumor and early reacting tissues. The repopulation is a function of time and potential doubling time. It is also assumed that repopulation is taking place at a constant rate.

In HDR, the treatment time is nearer to zero, all the 'β' damage is expressed, and hence, G = 1. On the other hand, if the time approaches infinity, G = 0, illustrating that all the 'β' damage are repaired. In the case of LDR, the treatment time is intermediate, repair begins to occur, but repair is incomplete. As a result, the expressed 'β' damage is decreased. Hence, G is a factor that compensate for incomplete repair called geometrical factor. It is an exponential function of treatment time (t), expressed in hours and SLD repair rate constant (μ):

$$G = \frac{2}{\mu t}\left[1 - \frac{1 - e^{-\mu t}}{\mu t}\right]$$

----- (19)

Thus, the LQ cell survival equation is re-written for LDR, incorporating G and replacing $d = R \times t$ where,

R is the LDR dose rate in Gy per hour and t is the LDR treatment duration in hours. Assuming $n = 1$ and ignoring repopulation correction:

$-\ln S = (\alpha d + \beta d^2)$

$-\ln S = (\alpha Rt + G \beta R^2 t^2)$

$$-\ln S = \alpha Rt + \frac{2}{\mu t}\left[1 - \frac{1-e^{-\mu t}}{\mu t}\right]\beta R^2 t^2$$

Taking Rt out and dividing by α:

$$\frac{-\ln S}{\alpha} = Rt\left[1 + \frac{2Rt}{\mu t (\alpha/\beta)}\left(1 - \frac{1-e^{-\mu t}}{\mu t}\right)\right]$$

$\text{BED}_{LDR} =$

$$\frac{-\ln S}{\alpha} = Rt\left[1 + \frac{2R}{\mu (\alpha/\beta)}\left(1 - \frac{1-e^{-\mu t}}{\mu t}\right)\right]$$

----- (20)

This is the final LDR equation, however the above equation can be approximated by leaving right side exponential term in the bracket:

$\text{BED}_{LDR} =$

$$Rt\left[1 + \frac{2R}{\mu (\alpha/\beta)}\right] = D\left[1 + \frac{2R}{\mu (\alpha/\beta)}\right]$$

----- (21)

where,

μ is the repair rate constant of the tissue and its unit is h^{-1}. If it is a permanent implant, then:

$$\text{BED}_{LDR} = R_0 \lambda \left[1 + \frac{1}{(\mu + \lambda)}\frac{R_0}{(\alpha/\beta)}\right]$$

----- (22)

where,

R_0 is the initial dose rate, and λ is the radioactive decay constant. The value of $\mu = 0.693/T_{1/2}$, in this, $T_{1/2}$ is the repair half-time of SLD in hours. The μ and repair half-time values for normal tissues are 0.46 h^{-1} and 1.5 h respectively. The corresponding μ and repair half-time values for tumor are 0.46–1.4 h^{-1} and 1.5–0.5 h respectively (Orton CG, 1990).

The repair half-time value for late reacting normal tissues including bladder and rectum is about 0.2–0.4 hours, if $\alpha/\beta = 2 - 4$ Gy, as reported in literature (Guerrero and Li). Roberts et al has suggested α/β value of 3 Gy and 10 Gy for late reacting normal tissues and tumor respectively. The tissue specific radiobiological parameters for gynecological brachytherapy are given in Table 10.2.

Table 10.2: Tissue specific parameters for gynecological brachytherapy (Thayalan K, 2002)

Tissue	α/β (Gy)	μ, h^{-1}	T_k (day)	k (Gy/day)
Point A	10.0	0.46–1.4	28	0.6
Rectum	3.87	0.46	0	0
Bladder	4	0.46	0	0

Worked example 5

In a HDR treatment of carcinoma (Ca) cervix a single fraction of 7 Gy is delivered. Calculate the BED value for tumor, bladder and rectum ($\alpha/\beta = 10$ Gy for tumor, 4 for rectum and 3.87 for rectum, ignore repopulation correction).

1. **Tumor:**
$$\text{BED}_{HDR} = 1 \times 7 \left(1 + \frac{7}{10}\right) = 11.9$$

2. **Bladder:**
$$\text{BED}_{HDR} = 1 \times 7 \left(1 + \frac{7}{4}\right) = 19.25$$

3. **Rectum:**
$$\text{BED}_{HDR} = 1 \times 7 \left(1 + \frac{7}{3.87}\right) = 19.66$$

Worked example 6

A LDR treatment is delivered to Ca. cervix patient with a dose rate of 1 Gy/h in 30 hours, to a total dose of 30 Gy. Calculate the tumor

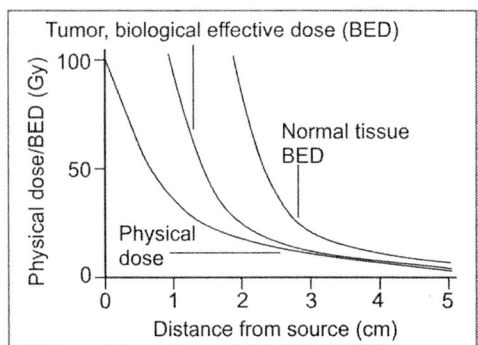

FIGURE 10.2: The variation of physical dose, tumor, biological effective dose (BED) (α/β = 10) and normal tissue BED (α/β = 3) with distance and dose rate from a brachytherapy source.

BED value for the treatment. Ignore repopulation and assume repair half-time is 1 h and μ = 0.693 h^{-1}, α/β = 10 Gy, G = 0.8.

Here, R = 1 Gy h^{-1}, t = 30 h, D = 30 Gy, and the equation 21 is given by:

$$BED_{LDR} = Rt \left[1 + \left(\frac{2R}{\frac{\alpha}{\beta}\mu} \right) \right]$$

$$BED_{LDR} = 1 \times 30 \left[1 + \left(\frac{2 \times 1}{10 \times 0.693} \right) \right] = 38.65 \text{ Gy}$$

The BED values of the tumor and normal tissues are always higher than that of the physical dose. The calculated BED also vary with distance and with dose rate from a radioactive source, similar to inverse square law (Fig. 10.2).

Role of RBE in Brachytherapy

There is no role of RBE in brachytherapy, since the conventional radioisotopes emits gamma rays of RBE unity. However, there are few radioisotopes used in LDR brachytherapy that have RBE > 1. For example, the RBE of I-125 is 1.45. This implies that I-125 require lesser dose to produce the same biological effect of 1 Gy X-ray dose. Hence, there is a need to account RBE in the LDR equation:

$$BED_{LDR} = R_o/\lambda \left[RBE_{max} + \frac{1}{(\mu + \lambda)} \frac{R_0}{(\alpha/\beta)} \right]$$

----- (23)

But, the issue is that RBE varies with dose rate and it is highest for low dose or low-dose rate. The RBE$_{max}$ refers to the RBE at zero dose or dose rate.

Conversion of LDR into HDR

The LQ model can be used to convert LDR into HDR in terms of BED. In doing so, the BED of LDR should be equal to BED of HDR. This model demonstrates that the HDR can be made equal to LDR, provided it is delivered in multiple fractions. This number of fractions depends on the tissue parameters such as:

- Repair rate constant
- Geometrical sparing.

In such situations, cancer site also play an important role on the conversions. The cervix is the most understood site, whereas others are yet to be established:

$$BED_{LDR} = BED_{HDR}$$

$$D \left[1 + \frac{2R}{\mu(\alpha/\beta)} \left(1 - \frac{1-e^{-\mu t}}{\mu t} \right) \right]$$

$$= n.d \left(1 + \frac{d}{\alpha/\beta} \right)$$

Multiplying by α/β, the equation becomes:

$$d^2 + d(\alpha/\beta) - \frac{D}{n}(\alpha/\beta).$$

$$\left[1 + \left(\frac{2R}{\mu \frac{\alpha}{\beta}} \right) \left(1 - \frac{1-e^{-\mu t}}{\mu t} \right) \right] = 0$$

By solving the above equation, the dose rate of HDR treatment with n number of fractions is given by:

$$d = \frac{-\frac{\alpha}{\beta} + \sqrt{(\alpha/\beta)^2 + \frac{4D}{n}(\alpha/\beta)\left[1 + \frac{2R}{\mu\frac{\alpha}{\beta}}\left(1 - \frac{1-e^{-\mu t}}{\mu t}\right)\right]}}{2} \quad \text{-----(24)}$$

where,

d is the dose per fraction in a HDR treatment schedule. There is no single dose that can satisfy all the three constraints, namely tumor control, early normal tissue reaction and late normal tissue reactions. However, compromise can be made in such situations, by opting the dose of tumor that will bring equivalent LDR local control.

To investigate the number of HDR fractions in gynecological tumor, Fowler has drawn the isoeffect curves with late normal tissue damage verses number of fractions (Fig. 10.3). He has compared the tumor effect of LDR with 70 Gy in 140 hours assuming the α/β = 3 and obtained a BED value for late normal tissue as 120. The horizontal lower dashed line refers to the highest BED of late normal tissue. The upper dotted line represents 5% higher BED to the normal tissue, which is acceptable in gynec brachytherapy. The solid lines refer to the percentage of tumor dose that the late normal tissue receives for various fractions. The minimum fractions that are required to subject the late normal tissue to 100% of tumor dose is 25–30. Similarly, the corresponding minimum fractions for 90% of tumor dose are 12–15. However, these fractions are not clinically executable in brachytherapy. Thus, normal tissue damage becomes the limiting factor in dose delivery in HDR treatments.

The minimum number of fraction required is 4–5 for 80% of the tumor dose. This matches well with late normal tissue damage and become the upper limit for treatment. Thus, if the dose to the late normal tissue is < 80% of the tumor dose, the HDR results are comparable to LDR. If the number of fraction decreases, the late normal tissue complication increases. Thus, some compromise is made in the above number (1 fraction) due to additional geometric sparing available in HDR, in the form of packing and retraction of normal tissues (bladder and rectum).

Worked example 7

A treatment schedule has 50 Gy external beam radiation therapy (EBRT) followed by 30 Gy LDR brachytherapy (Dose rate = 1 Gy per hour, 30 hour). What is the equivalent HDR treatments, if so the number of fractions?

$$BED_{EBRT} = 25 \times 2 \left(1 + \frac{2}{10}\right) - 0.6(35 - 28)$$

$$= 55.8$$

$BED_{LDR} = 38.65$ (refer worked example 6)

$$\text{Total BED}_{EBRT} + BED_{LDR} = 55.8 + 38.65$$

$$= 94.45$$

To achieve the total BED of 94.45 Gy, 3 fractions of either 7 Gy or 7.5 Gy per fraction should be used. The calculated values of BED are given in the Table 10.3.

To conclude the total EBRT + LDR brachytherapy BED exactly lies in between 3 HDR brachytherapy fractions of size 7 and 7.5 Gy. Mathematically, the 7.5 Gy HDR

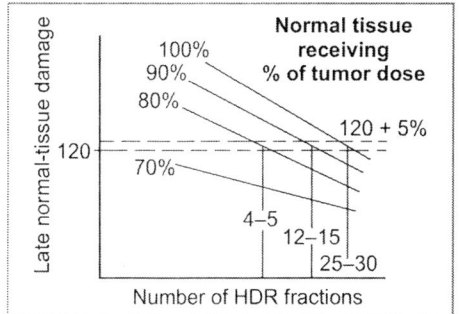

FIGURE 10.3: Relation between late normal-tissue damage and number of high-dose rate (HDR) fractions that gives equivalent tumor effects to low-dose rate (LDR) 70 Gy in 140 hours (α/β = 3, BED = 120).

Table 10.3: The biologically effective dose (BED) values of tumor, bladder and rectum in a external beam radiation therapy (EBRT) + low-dose rate (LDR) equivalent EBRT + high-dose rate (HDR) treatment schedule for carcinoma cervix (α/β = 10 for tumor, 4 for bladder, 3.87 for rectum and μ = 0.693 h^{-1}).

Modality	Tumor BED (Gy)	Bladder BED (Gy)	Rectum BED (Gy)
EBRT	55.8	75	75.8
LDR	38.65	62.6	63.7
Total = EBRT+ LDR	94.45	137.6	139.5
HDR at 7 Gy/fx	11.9	19.25	19.66
EBRT + 3HDR at 7 Gy/fx	91.5	132.75	134.78
HDR at 7.5 Gy/fx	13.1	21.56	22
EBRT + 3 HDR at 7.5 Gy/fx	95.1	139.68	141.8

fraction seems to be more toxic to bladder and rectum than 7 Gy HDR fraction. However, HDR fractions involves interval between fractions, of the order of week, hence, there is decay of biological effect. Hence, it is recommended to use 7.5 Gy per fraction in 3 fractions as equivalent to traditional LDR treatment. Any reduction in the dose per fraction or the number of fractions certainly leads to tumor recurrence. Similarly, increasing the dose per fraction > 7.5 Gy and reducing the number of fractions less than 3 may also cause recurrence with increased normal tissue toxicity.

LQ Model for Combined Radiotherapy

If a treatment is planned with EBRT followed by HDR brachytherapy in multiple fractions, then the cumulative BED for the whole combined treatment is given by:

$$BED_{Total} = BED_T - 0.5G_1 + BED_{HDR1} - 0.5G_2 + BED_{HDR2} + ... \quad \text{-----(25)}$$

where,
- T = External beam radiotherapy
- G_1 = Gap in days between EBRT and HDR1
- G_2 = Gap in days between HDR1 and HDR2.

Worked example 8

A hospital uses a combined radiotherapy treatment protocol of 50 Gy EBRT in 5 weeks in 2 Gy per fractions, followed by 7.5 Gy, 3 HDR fractions and each treatment is spaced 1 week apart for carcinoma cervix. Calculate the cumulative BED for the treatment.

1. **EBRT:**
   Here,
   N = 25
   d = 2 Gy
   T = 35 days,
   G1 = G2 = G3 = 7 days
   The LQ equation is:
   $$BED = 25 \times 2\left(1 + \frac{2}{10}\right) - 0.6(35 - 28) = 55.8$$

2. **HDR:**
   $$BED = 7.5 \times 1\left(1 + \frac{7.5}{10}\right) = 13.1 \text{ Gy}$$

3. **EBRT + HDR:**
   $$BED_{Total} = BED_T - 0.5G_1 + BED_{HDR1} - 0.5G_2 + BED_{HDR2}$$
   $$= (55.8 - 0.5 \times 7) + (13.1 - 0.5 \times 7) + (13.1 - 0.5 \times 7) + 13.1$$
   $$= 52.3 + 9.6 + 9.6 + 13.1 = 84.6 \text{ Gy}$$

Worked example 9

A patient is implanted interstitially for tongue cancer and offered 5 HDR fractions, 2 fractions per day, with a fraction size of 3.5 Gy. Calculate the BED value and find out the LDR equivalent dose (assume $\alpha/\beta = 13.4$).

$$BED_{HDR} = 5 \times 3.5 \left(1 + \frac{3.5}{13.4}\right) = 22$$

The total tumor BED = 22 Gy, which is equivalent to LDR dose of 20 Gy delivered at the rate of 0.5 Gy/h, for 40 hours (Table 10.4).

The LQ model hold good only for dose per fraction up to 6 Gy. For higher dose per fraction, it tends to overestimate the magnitude of cell kill, since the survival curve become linear again at higher dose per fraction. In addition, there will be interpatient heterogeneity in radiobiological parameters that are used in LQ model. Hence, there are certain important issues yet to be solved in the combined BED that includes:

1. Exact values of radiobiological parameters are unknown.
2. Variation of BED over distance due to high-dose gradients and its dependence on dose rate.

Some amount of clarity on the above issues will be dealt in the following paragraphs.

INFLUENCE OF RADIOBIOLOGICAL PARAMETERS

Generally, tumor grows slowly before treatment than afterwards, due to poor oxygenation. This results in low growth rate and high cell loss. After radiation or surgery, cell kill occurs, the remaining cell become better oxygenated, resulting increase in tumor growth rate. The above events are influenced by various radiobiological parameters as given below:

1. Radiosensitivity parameter (α).
2. Fractionation sensitivity parameter (α/β).
3. Pretreatment doubling time (T_{eff}).
4. Cell doubling time in the absence of cell loss (T_{pot}).
5. Repair half-time ($T_{1/2}$) and number of clonogenic cells at the time of diagnosis (N_0).

These radiobiological parameters vary with different types of tumor and its sensitivity (Table 10.5). The value of α is 0.3 Gy^{-1} for medium tumor sensitivity. The corresponding

Table 10.4: Low-dose rate (LDR) equivalent high-dose rate (HDR) dose at 2 fractions per day for interstitial implants in head and neck brachytherapy ($\alpha/\beta = 13.4$, $\mu = 0.693$ h^{-1})

LDR dose (Gy), at 0.5 Gyh^{-1}	Biologically effective dose (BED) (Gy)	External beam radiation therapy (EBRT) dose (Gy) N* × d†	HDR equivalent protocol at 2 fractions per day	
			Dose/Fraction (Gy)	Number of fractions
10	11.0	5 × 2	3	3
			4.2	2
15	16.6	7 × 2	3.4	4
			4.2	3
20	22.0	9 × 2	3.5	5
			4.2	4
25	27.5	10 × 2	3.6	6
			4.5	5
30	33.2	14 × 2	3.7	7
			4.5	6

*N, number of fractions; †d, dose per fraction.

Table 10.5: Radiobiological parameter for various tumor sites (Ruth Wyatt, 2011)

Radiobiological parameter	Head and Neck (SCC)*	Prostate	Breast	Cervix
α, Gy^{-1}	0.3	0.15–0.155	0.27	0.3
α/β, Gy	10.5	3.1 or 1.5	4	10
T_{eff}, days	45 (before surgery) 12 (after surgery)	1148	26 (all tumors) 14 (rapidly growing)	119
T_{pot}, days	2.3	42	26 (all tumors) 14 (rapidly growing)	6.8
$T_{1/2}$, days	–	–	–	0.5
N_o	5×10^7 (T2, T4) 1×10^5 (T1, no surgery) 5×10^3 (postsurgery)	3×10^6 to 26×10^9	8×10^3 (postsurgery)	2×10^8

*Squamous cell carcinoma

fractionation sensitivity parameter, α/β is 10 for tumors, which are relatively insensitive to fractionation. However, some tumors, such as breast may have lesser α/β ratio that of close to normal tissues.

The tumor doubling time during treatment is similar to the tumor cell doubling time in the absence of cell loss (T_{pot}). Hence, T_{pot} may be considered as the representative parameter of tumor doubling time during treatment for a given tumor type. However, it is not a good predictor of radiotherapy tumor control for individual patients. There is a existence of lag period before on set of accelerated repopulation and it is found to be 20–32 days for head and neck cancer. The lag time for other tumors are yet to be investigated.

There is also uncertainty and also interpatient variation in many radiobiological parameters. Investigations reveal that interpatient heterogeneity is dominated by heterogeneity in the radiosensitivity parameter. For example, 'α' value varies by 20–50%, hence, use of distributed values of α ($\pm 2\sigma$) and T_{pot} is suggested. There is also variation in the estimated number of clonogenic cells in a typical tumor. The clonogenic cell density of breast and cervix cancer is about 10^7 cm^{-3} at the time of diagnosis. Cervix cancer is assumed to have medium radiosensitivity (α = 0.3 Gy, α/β 10 Gy) and its heterogeneity is about 30%, which shows good response to radiation. However, prolongation of OTT may have deleterious effect on outcome.

DOSE-FRACTIONATION EFFECTS IN GYNECOLOGICAL BRACHYTHERAPY

Treatment of gynecological cancer consists of major proportion of brachytherapy that too with high-dose rate brachytherapy. Though HDR is fractionated, to equate to LDR or MDR brachytherapy, the exact number of HDR fractions are not well understood.

Equivalent Dose

Equivalent dose is difficult to calculate the tumor control probability in all cancers. This is true especially in carcinoma cervix, where the dose distribution is inhomogeneous, point A is not the representative of tumor and often chemotherapy is administered to the patient. Both LDR and HDR repopulation is ignored, based on the assumption that the treatment is completed within the specified OTT. However, BED calculation should incorporate correction for repair component that is taking place during LDR treatments.

Hence, the estimated BED may not represent exact treatment doses. This requires alternate methods of calculation.

The equivalent dose (EQD) concept was proposed by Preterite (1999), so that the total dose delivered is equated to standard 2 Gy EBRT equivalent. This is usually known as EQD, which is the ratio of the BED and relative effectiveness:

$$EQD_2 = \frac{BED}{\text{Relative effectiveness}}$$

$$EQD_2 = \frac{BED}{\left(1 + \dfrac{2}{\alpha/\beta}\right)}$$

if the value of α/β is 10, then

$$EQD_2 = \frac{BED}{\left(1 + \dfrac{2}{10}\right)} = \frac{BED}{1.2} \quad \text{-----(26)}$$

This equation is valid for tumor; similarly one can derive for late reacting tissues such as bladder and rectum by using their α/β value respectively. The BED for low- and medium-dose rate treatments depends on the repair half-time of the tissue, and α/β value of the tissue. These are not well-known for tumor and normal tissues as on date. In general, α/β value of 3 Gy for normal tissue, 10 Gy for tumor, $T_{1/2}$ = 0.5 h for tumor and 1.0 h for normal tissue is used.

Both long and short repair half-time have been suggested for rectum and bladder, while comparing LDR and HDR. Guerrero M and Li (2006) have suggested the most likely value for rectum and bladder is about 0.2–0.4 h, if α/β = 2 – 4. Roberts et al (2004) have estimated the repair half-time as 0.25 h and 0.5 h for tumor and normal tissue respectively. They assumed α/β = 3 Gy and 10 Gy for normal tissue and tumor respectively and also estimated sparing factor for normal tissue. However, the α/β value and the repair half-time for the rectum is estimated as 5.4 Gy and 1.0 h respectively (Brenner D, 2004).

Change of BED Over Distance and the Effect of Dose-fractionation

Brachytherapy in gynecology is associated with high-dose gradients. Hence, the HDR and LDR are equivalent only at one point in space. The BED of the tumor and normal tissue changes over distance from the source. This effect is larger for fractionated HDR, since there is larger sparing of normal tissue (Fig. 10.4A). This figure is based on $t_{1/2}$ = 0.5 h and α/β =10 Gy for tumor. The BED of LDR

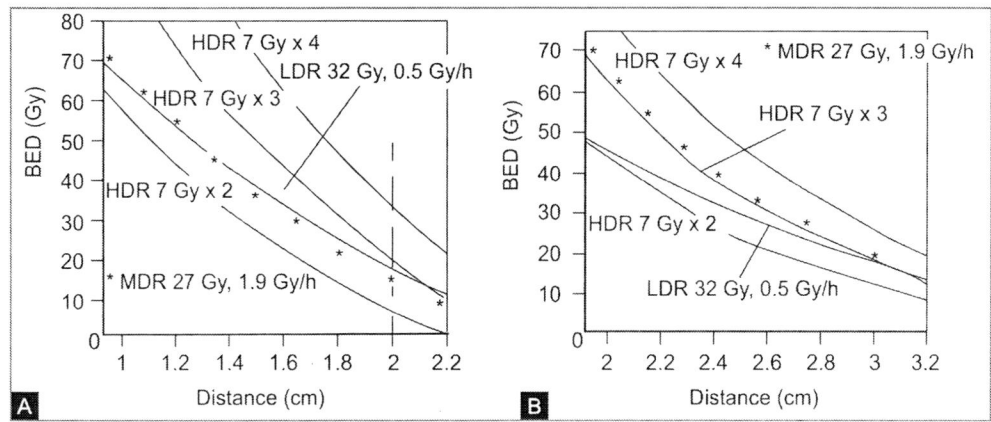

FIGURES 10.4A and B: Change of biologically effective dose (BED) with distance for high-dose rate (HDR), medium-dose rate (MDR) and low-dose rate (LDR) brachytherapy. **A.** Variation of tumor BED ($t_{1/2}$ = 0.5 h, α/β = 10 Gy); **B.** Variation of normal tissue BED ($t_{1/2}$ = 1.0 h, α/β = 3 Gy).

treatment of 32 Gy with 0.5 Gy/h dose rate curve lies in between 7 Gy HDR fractions of 3 and 2. It is slightly lesser than the HDR 7 Gy × 3 at the prescription point of 2 cm. It can be shown that it is also equivalent to 27 Gy of MDR with dose rate 1.9 Gy/h. However, at lesser distance of less than 2 cm, the fractionated HDR BED increases faster whereas the LDR increases slowly. The quadratic term is small in the LDR equation due to protracted treatment times, which is usually influenced by dose per fraction and dose rate. However, the BED of HDR remains the same, whereas that of LDR and MDR is lower, due to more repair during treatment. Similar conclusion can be drawn for repair half-time of $t_{1/2} = 0.25$ and $\alpha/\beta = 10$ Gy.

The above discussion can be extended to normal tissue BED ($\alpha/\beta = 3$ Gy, $t_{1/2} = 1$ h) over distance from the source (Fig. 10.4B). The LDR treatment of 32 Gy with 0.5 Gy/h dose rate is approximately equivalent to 2 fractions of 7 Gy HDR treatments at the prescription point of 2 cm distance. It is lesser than the 27 Gy of MDR with dose rate 1.9 Gy/h, which is equal to 3 fractions of 7 Gy HDR treatments. However, at distances greater than 2 cm, the dose rate fall off and obey inverse square law. The fractionated BED of HDR decreases faster than the BED of LDR. The LDR component decreases slowly because the quadratic term is small as discussed earlier. Similar conclusions can be made for $\alpha/\beta = 5$ Gy, $t_{1/2} = 1$ h also, but the total BED might be lower.

GAPS IN RADIATION TREATMENT

Gaps may arise during the course of radiation treatment, due to machine failure or radiation reaction on patient or illness, etc. This may be planned and unplanned interruptions. Interruptions have greater impact on radiation treatment. This should be overcome in order to maintain the same level of tumor control. Generally, radiation oncologists either ignore the missed fractions or add the missed fractions at the end of treatment course. Both have implications on tumor control. The former option will reduce the total dose and bring down tumor control. The latter option will extend the OTT and allows cell repopulation, leads to significant loss in tumor control, especially in rapidly proliferating tumor.

This includes chemotherapy, brachytherapy and external beam therapy. Although extended OTT will spare the acute normal tissue reactions, the risk of late effects may not be reduced. A reduction in cure probability of 1–2% per day of extension of OTT is reported in squamous cell carcinoma (SCC) in head and neck cancer. The corresponding value for cervix and lung is 0.7–1% and 1.3–2% per day of prolongation respectively. However, if long interruption comes at the end of treatment, extension of overall treatment is unavoidable. There will be loss in therapeutic index.

Alternatively, increase of dose per fraction can be explored within the planned treatment time. Though repopulation is effectively controlled, the patient may feel the adverse effect of high dose per fraction. The best way is to add fractions at week end or giving multiple fractions per day, keeping the OTT remains the same. Overall, patients receiving radical treatment should not have interruptions during the duration of treatment. This is very important in rapidly proliferating tumors. If the interruption is due to machine failure, alternate arrangements should be made to treat the patient in other machines. Every hospital should have specific policy or guidelines for the management of interruptions. In general, the interruption management methodologies are either by retaining the OTT or by delivering extra fractions over extended dose per fraction. In the later, the reduction in biological dose should be compensated per day of prolongation. Hence, the k factors become crucial and it is 0 for normal tissue. The k value for tumor is 0.9 Gy/day, ≤ 0.5 Gy/day and 0.3 Gy/day respectively for head and neck, cervix and breast.

Gap Correction Procedure by Retaining Overall Treatment Time

1. Calculate the planned overall tumor BED and normal tissue BED.
2. Calculate the delivered tumor BED and normal tissue BED.
3. Calculate the remaining normal tissue BED yet to be given, in order to equate the planned normal tissue BED.
4. Calculate the treatment days and number of fractions remaining, in order to equate the planned OTT.
5. Calculate the dose per fraction from the remaining normal tissue BED.
6. Calculate the tumor BED additionally given with above dose per fraction and estimate the overall tumor BED achieved.
7. If the overall tumor BED is too low, a slight increase in normal tissue BED can be accepted.
8. Treatment at week end or delivering multiple fractions per day may be explored.

Worked example 10
Head and neck cancer patient is planned with a standard external beam treatment of 60 Gy, 2 Gy per fraction delivered in 30 fractions in 40 days. He developed acute reactions after 25 fractions and is given 5 days rest and treatment resumed on day 39. Calculate the dose per fraction to complete the treatments in 2 fractions, ignoring the effect of OTT (assume tumor α/β = 10 Gy, T_k = 28 days and k = 0.9 Gy/day, and late reacting normal tissue α/β = 3.0 Gy and k = 0 Gy/day).

Planned tumor BED = 72 Gy_{10} and normal tissue BED = 99.6 ≈ 100 Gy_3 (excluding repopulation correction).

Delivered tumor BED = 60 Gy_{10} and the normal tissue BED = 83 Gy_3.

The remaining normal tissue BED to be given = 100 − 83 = 17 Gy_{10} and the fraction size is obtained as follows:

$$17 = 2 \times d \left(1 + \frac{d}{3}\right)$$

d = 3.7 Gy, delivered in 2 fractions in 2 days:

Tumor BED additionally given

$$= 2 \times 3.7 \left(1 + \frac{3.7}{10}\right) = 10.1 \text{ Gy}$$

The overall delivered tumor BED = 60 + 10.1 = 70.1 Gy, this is lesser than the planned tumor BED of 72 Gy. The 10.1 Gy is equal to 4.2 fractions of 2 Gy, which raise the total fraction into 29.2 (25 + 4.2). This is equivalent to 58.4 Gy (29.2 × 2 Gy) in 2 Gy fractions over 40 days. Since the delivered tumor BED is inferior to the planned treatment regimen, some compromise can be made in normal tissue tolerance, by delivering 3 fractions of 3 Gy instead of 2 fractions of 3.7 Gy each. This should be delivered as one fraction on day 39 and two fractions on day 40 with sufficient interval of time:

Tumor BED additionally given

$$= 3 \times 3 \left(1 + \frac{3}{10}\right) = 11.7 \text{ Gy}$$

The overall delivered tumor BED = 60 + 11.7 = 71.7 Gy, this is exactly equal to planned tumor BED of 72 Gy. The 11.7 Gy is equal to 4.9 fractions of 2 Gy, which raise the total fraction into 29.9 (25 + 4.9). This is equivalent to 59.8 Gy (29.9 × 2 Gy) in 2 Gy fractions over 40 days.

Gap Correction Procedure by Delivering Extra Fractions Over an Extended Overall Treatment Time

1. Calculate the planned overall tumor BED and normal tissue BED.
2. Calculate the delivered tumor BED and normal tissue BED.
3. Estimate the missing fractions and simply extend the treatment time to deliver the missing fractions either with or without week end treatment.

4. If the tumor BED is lesser than the planned tumor BED, explore the possibility of multiple fractions per day, with reduced OTT.
5. Calculate the remaining tumor BED and estimate the dose per fraction to be delivered. Find the overall tumor BED, if it is too low, a slight increase in normal tissue BED can be accepted.
6. Calculate the number of fractions required for treating at week end or delivering multiple fractions per day.

Worked example 11

Head and neck cancer patient is planned with a standard external beam treatment of 60 Gy, 2 Gy per fraction delivered in 30 fractions in 40 days. He developed acute reactions after 25 fractions and is given 5 days rest and treatment resumed on day 39. Calculate the dose per fraction to complete the treatments in 5 fractions or extending the treatment time (assume tumor α/β = 10 Gy, T_k = 28 days and k = 0.9 Gy/day, and late reacting normal tissue α/β = 3.0 Gy, and k = 0 Gy/day).

Planned tumor BED = 61.2 Gy_{10} and normal tissue BED = 100 Gy_3 (including OTT of 40 days and repopulation correction).

Delivered tumor BED = 51 Gy_{10} and the normal tissue BED = 83 Gy_3 (including OTT of 38 days (33 + 5) and repopulation correction).

Option 1: The missing fractions are 5, simply extend the treatment for 5 days with one fraction per day with week end treatment. The OTT become 43 (38 + 5), the tumor BED is 58.5 Gy_{10} and the normal tissue BED is 100 Gy_3.

Option 2: The missing fractions are 5, simply extend the treatment for 5 days with one fraction per day without week end treatment. The OTT become 45 (38 + 7), the tumor BED is 56.7 Gy_{10} and the normal tissue BED is 100 Gy_3.

Option 3: The above tumor BED is lesser than the planned tumor BED. Hence, the missing 5 fractions can be given in 2 days with 2 fractions on day 39 and 3 fractions on day 40 with 6 hours interval. The OTT become 40 (38 + 2), the tumor BED is 61.2 Gy_{10} and the normal tissue BED is 100 Gy_3.

Option 4: The remaining tumor BED = 61.2 – 51 = 10.2 Gy_{10}. This can be achieved by delivering:
1. Four fractions at 2.1 Gy per fraction in 4 days.
2. Three fractions at 2.7 Gy per fraction in 3 days.
3. Two fractions at 3.7 Gy per fraction in 2 days.

The corresponding BED achieved for normal tissue is 14.3 Gy_3, 15.4 Gy_3 and 16.5 Gy_3 respectively. The overall normal tissue BED achieved is 97.3 Gy_3, 98.4 Gy_3 and 99.5 Gy_3 respectively.

Option 5: The remaining tumor BED = 61.2 – 51 = 10.2 Gy_{10}. This can be achieved by delivering 2 fractions per day at the rate of 2 Gy per fraction with weekend treatments. One has to calculate the number of fractions required to satisfy remaining the tumor BED:

$$10.2 = 2 \times n \left(1 + \frac{2}{10}\right) - 0.9 \times \frac{n}{2}$$

The number of fraction, n is found to be 5.2 ≈ 5. Hence, the recommended treatment schedule is 5 fractions of 2 Gy per fraction, twice daily in 3 days. The corresponding BED for tumor and normal tissue achieved is 9.75 Gy_{10}, and 16.7 Gy_3 respectively. The overall tumor and normal tissue BED achieved is 60.8 Gy_{10}, and 99.7 Gy_3 respectively. This exactly satisfies the planned treatment schedule of 61.2 Gy_{10} and 100 Gy_3 respectively.

Gap Correction for Incomplete Repair

The above said corrections are applicable only to clinical situations where complete repair is accounted. But, there are situations where repair time is longer, leading to incomplete repair, e.g. spinal cord. Hence, there is a need of radiobiological equation that incorporate incomplete repair, while attempting multiple fractions. As per the reciprocal time model, (Dale RG et al, 1999), the BED is given by:

$$BED = N \times d \left(1 + \frac{(1+h)d}{\alpha/\beta} \right) \quad \text{-----(27)}$$

where,

$$h = \frac{2}{N} \sum_{i=1}^{N-1} \frac{N-i}{1+i+(f/t\frac{1}{2})}$$

f is the time between fractions, N is the number of fractions and d is the dose per fraction. This accounts the slowing down of repair rate with increasing time after irradiation. The suggested weighted repair half-time is 2.56 h and α/β is 2 Gy (Ang KK et al, 1992) for spinal cord. However, this is valid for constant dose per fraction and constant interfraction interval.

BIBLIOGRAPHY

1. Colin G Orton. Fractionation: Radiobiologic principles and clinical practice. In: Faiz M Khan (Ed). Treatment Planning in Radiation Oncology, 2nd edition. Philadelphia, USA: Lippincott Williams & Wilkins; 2007.
2. Colin G Orton. Radiobiology. In: Subir Nag MD (Ed). High Dose Rate Brachytherapy. Armonk, NY: Futura Publications Company Inc; 1994.
3. Colin G Orton. Radiobiology. In: Subir Nag MD (Ed). Principles and Practice of Brachytherapy. Armonk, NY: Futura Publications Company Inc; 1997.
4. Eric J Hall, Amato J Giaccia. Radiobiology for the Radiologist, 7th edition. Philadelphia, USA: Wolters Kluwer/Lippincott Williams & Wilkins; 2012.
5. Loredana G Marcu. Eva Benzak. Recent advances and Research Updates: Medical Physics and Radiobiology. Trivandrum, India: Researchman Publishers Pvt Ltd; 2011.
6. Mayles P, Nahum A, Rosenwald JC. Handbook of Radiotherapy Physics-Theory and Practice. New York: Taylor & Francis; 2007.
7. Phillip M Devlin. Brachytherapy Applications and Techniques. Philadelphia, USA: Lippincott Williams & Wilkins; 2007.
8. Sanjay S Supe. An Overview of Time Dose Fractionation Models; NSD to BED. Pol J Med Phy Eng. 2006;12(4):165-201.
9. Thayalan K. Physical and dosimetric studies of high dose rate brachytherapy system with clinical correlation, in carcinoma of the uterine cervix: PhD thesis. Chennai, India: The Tamil Nadu Dr MGR Medical University Library; 2002.

Radiation Safety and Protection

CHAPTER 11

EQUIVALENT DOSE AND EFFECTIVE DOSE

The hazards of radiation were realized in the beginning of this century, soon after the discovery of X-rays. In 1921, the British X-ray and Radium Protection Committee recommended maximum tolerance dose. Radiation exposure limits were introduced by the International Commission on Radiological Protection (ICRP) in 1928. The ICRP reports form the basis for many national and international radiation protection programs. In India, the radiation protection was started with the formation of Directorate of Radiation Protection (DRP) at the Bhabha Atomic Research Centre, in 1963. Atomic Energy Regulatory Board (AERB) established in 1983 is the competent authority, which controls over the safe use of radiations both in medicine and industry.

Exposure to radiation may be whole-body exposure or partial body exposure. In addition, it can be uniform or non-uniform exposure. Radiosensitivity of tissues in human body varies toward radiation exposure. Individual tissues contribute differently to the total detriment of health of the exposed person, depending upon the seriousness of the damage and its curability. For quantitative analysis, ICRP has recommended two parameters, namely:
- Equivalent dose
- Effective dose.

Equivalent Dose

Hence, all radiation cannot cause the same biological damage per unit dose. In order to account the above factor, ICRP (Report, 1990) has introduced radiation weighting factor (W_R). It will modify the dose to reflect the relative effectiveness of the type of radiation causing biological damage. The product of the absorbed dose and radiation weighting factor is called equivalent dose (H):

$$H = D \times W_R \quad \text{-----} (1)$$

where,

D is the absorbed dose and W_R is the weighting factor for the radiation type. The weighting factor is 1 for X-rays, gamma (γ)-rays and electron of all energies. High-LET radiation may cause higher biological effect, hence has higher radiation weighting factor. The radiation weighting factors for various radiations are given in the Table 11.1.

The SI unit of equivalent dose is Sievert (Sv) and 1 Sv = 1 J/kg. Rem is the special unit of equivalent dose, used early, when absorbed dose is measured in rad. Rem is the short form of radiation equivalent men and 100 rem is equal to 1 Sv. In practice, millisievert (mSv) is mostly used as follows:

1 Sv = 1,000 mSv or 100 rem

100,000 mrem = 1,000 mSv (1 rem = 1,000 mrem)

Hence, 100 mrem = 1 mSv or 100 mR = 1 mSV, if f, the rad-Roentgen conversion factor, is unity.

Table 11.1: Radiation weighting factors (W_R)

Radiation types	W_R
Photons	1
Electrons and muons	1
Protons and charged pions	2
Alpha particles, fission fragments, heavy ions	20
Neutrons	Continuous curve as a function of neutron energy (2.5–20)

Source: The 2007 Recommendations of the International Commission on Radiological Protection. ICRP publication 103. Ann ICRP. 2007;37(2-4):1-332.

TABLE 11.2: Tissue weighting factors (W_T)

Tissues	W_T	Total contribution
Bone marrow, colon, lung, stomach (4 tissues)	0.12	0.48
Breast (1)	0.12	0.12
Remainder*	0.12	0.12
Gonads (1)	0.08	0.08
Bladder, esophagus, liver, thyroid (4)	0.04	0.16
Bone surface, skin (2)	0.01	0.02
Salivary gland, brain (2)	0.01	0.02
Total		1.0

Source: The 2007 Recommendations of the International Commission on Radiological Protection. ICRP publication 103. Ann ICRP. 2007;37(2-4):1-332.

*Remainder tissues: Adrenals, gallbladder, heart, kidneys, lymphatic nodes, muscle, oral mucosa, pancreas, prostate, small intestine, spleen, thymus, uterus, extrathoracic region.

Effective Dose (H_T)

To account for the variation of radiosensitivity of different tissues and the nonuniformity of radiation exposure, ICRP has established tissue weighting factors (W_T) as given in the Table 11.2. The weighting factor of a particular tissue or organ has the risk of stochastic effects being induced in the organ when singly irradiated, compared to the total risk of inducing stochastic effects if the same radiation dose is received by the whole body. The sum of the products of the equivalent dose to each tissue irradiated (H_T) and the corresponding weighting factor of tissue is called effective dose (E):

$$E = \Sigma W_T \times H_T \quad \text{-----(2)}$$

where,

W_T is the weighting factor of tissue, T and H_T is the mean equivalent dose received by the tissue T. This quantity (E) expresses the overall measure of health detriment associated with each irradiated tissue as a whole-body dose and considers the radiosensitivity of each irradiated tissue. It is used to evaluate the probability of stochastic effects at low doses.

It was understood that testis and ovaries are the most radiosensitive tissues as per ICRP (1990). However, bone marrow and breast tissue are the most radiosensitive organs as per ICRP (2007). Organ of higher sensitivity carries a higher risk for a given dose. The sum of the weighting factors is unity. The unit of effective dose is also Sievert (Sv). The tissue weighting factors are developed from a reference population, having equal number of sexes and wide range of age.

Committed Equivalent and Effective Dose

The committed equivalent dose and committed effective dose terminologies are used in internally deposited radionuclides. Following an intake of radioactive material, there is a period during which the deposited material gives rise to equivalent doses in the tissues of the body on a time-dependent rate. The time integral of the equivalent dose is called committed equivalent dose. Similarly, the committed effective dose can also be determined. Depending upon the biokinetics of the radionuclides, different organs may receive varying amounts of exposure.

Collective Equivalent and Effective Dose

The collective equivalent dose and collective effective dose are applicable to a group of persons or population as a whole. The number of persons exposed from a radiation source should be known and then the dose value is arrived at by summing up their individual doses. Alternatively, the product of average individual dose and the number of persons exposed will also give the collective dose. Their unit is person Sievert. It is believed that the above two parameters represent the total consequences (detriment) of the exposure accruing to the population group.

SOURCES OF RADIATION

Every day we are exposed to radiation from various sources. The sources of radiation are classified into (Fig. 11.1):
- Natural radiation sources
- Enhanced natural sources
- Artificial radiation sources (manmade)
- Occupational exposures.

The annual average per capita total effective dose equivalent by ionizing radiation is about 3.6 mSv [Table 11.3, International Atomic Energy Agency (IAEA)]. About 82% of the above exposure (3 mSv) arise from naturally occurring sources, 18% (0.6 mSv) arise from artificial radiation sources in which medical X-rays is the major contributor (58%). Background radiation involves both natural and manmade low level radiation exposure to all members of the public. This will vary with region; Kerala and Brazil have high background levels of radiation (> 10 mSv/year).

Table 11.3: Mean annual radiation exposure per person by ionizing radiation

Sources	Mean dose (range) mSv
Cosmic rays	0.4 (0.3–1.0)
Terrestrial	0.5 (0.3–0.6)
Inhalation (radon)	1.2 (0.2–10)
Ingestion	0.3 (0.2–0.8)
Medical	0.4 (0.04–1.0)
Nuclear testing	0.15
Nuclear power production	0.0002
Total	2.95

Source: International Atomic Energy Agency (IAEA), 2000.

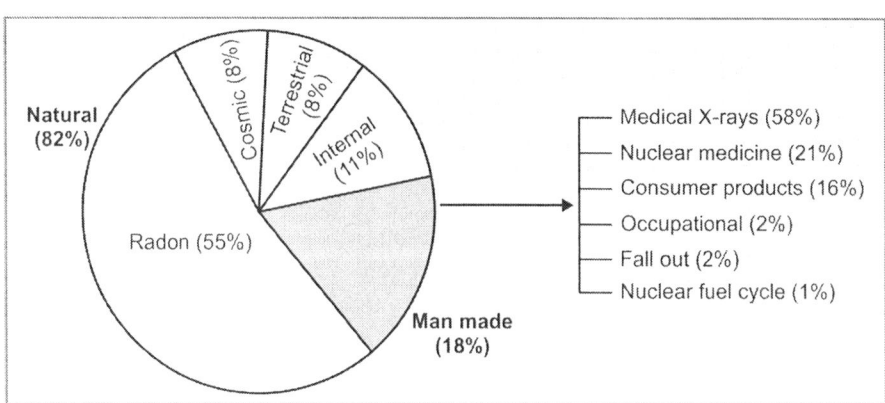

FIGURE 11.1: Different sources of radiation and their contribution in percentage

Natural Radiation Source

The natural radiation source includes:
- Cosmic rays
- Terrestrial radionuclides
- Internal radioisotopes.

Cosmic rays are outer world radiations comprising of energetic protons and alpha particles. They are divided into primary and secondary cosmic rays. The primary cosmic rays are protons collide with atmosphere, producing showers of secondary particles (electrons, muons) and electromagnetic radiations. Cosmic exposures increase with altitudes, it is doubling in every 1,500 m and it is greater at earth poles than at equator. An air traveler is exposed to increased level of cosmic rays. Travelling in a transcontinental flight may yield about 10 µSv per hour. Structures provide some protection against cosmic rays, and hence, the indoor effective dose is 20% lesser than outdoor.

Terrestrial radionuclides have been present since the formation of earth. They mainly contribute in the form of external exposure, inhalation and ingestion. U-238 and Th-232 give both external and internal exposure. The biggest contributor of terrestrial radiation is Rn-222 of the order of 1.2 mSv per year. Radon is the daughter product of naturally occurring U-238 in the soil. Hence, inhalation exposure is due to Rn-222. Radon's get attached to aerosols and are deposited in the lungs. They may irradiate bronchial mucosa and may induce bronchogenic cancer. Internal radionuclide includes K-40 and C-14, which are present in the human body. The main contributor is K-40, which emits β- and γ-rays.

Enhanced Natural Sources

Enhanced natural sources consist of building material, mining and agriculture, combustible fuel, consumer products such as tobacco and radon gas dissolved in domestic water supply. Building materials (brick, concrete and granite) consists of uranium, thorium and potassium. Mining and agricultural activity contribute to a lesser level by fertilizers (uranium, thorium decay products and K-40). Combustible fuels include coal, natural gas and consumer products include smoke alarms (americium-241), gas lantern mantles (thorium), dental prostheses, certain ceramics, optical lenses (uranium).

Artificial Sources

The artificial or manmade sources of radiation include medical exposure, radioactive fallout, nuclear power and occupational exposure. Majority of medical exposure is from medical X-rays (fluoroscopy and computed tomography) and the next contributor is the nuclear medicine. Both produce an annual average effective dose equivalent of 540 µSv per year, which is about 79% of artificial radiation.

Consumer products such as tobacco, the domestic water supply, building materials and to a lesser extent smoke detectors, televisions and computer screens, account for 16% of the artificial radiation exposure.

The occupational exposure arises from uranium mining (12 mSv per year) nuclear power operations, medical diagnosis and therapy, aviation and research, non-uranium mining and application of phosphate fertilizers. It contributes about 2% of the artificial radiation exposure. The uranium miners, nuclear power operators, air crews, X-ray technologists and radiologists receive an annual effective dose equivalent of 12, 6, 1.7, 1.0 and 0.7 mSv per year respectively.

Atmospheric testing of nuclear weapons give out Carbon-14 and other radionuclides, including H-3, Mn-54, Cs-136, 137, Ba-140, Ce-144, plutonium and transplutonium elements. The contribution from nuclear power production is about 1% of artificial radiation and the most significant contributor is Carbon-14.

PRINCIPLES OF RADIATION PROTECTION

The aim of radiation protection should be to prevent deterministic effects and minimize the probability of stochastic effects to levels deemed to be acceptable. This could be achieved:
1. By setting limits well below threshold dose to deterministic effects.
2. The probability of stochastic effects could be reduced by limiting exposures as low as reasonably achievable (ALARA).

The whole radiation protection philosophy can be summarized as follows:
1. **Justification of practice:** No practice involving radiation exposures shall be adopted unless it produces a net positive benefit.
2. **Optimization:** Every effort shall be taken to reduce the dose as low as reasonable achievable, considering the clinical, social and economic factors.
3. **Dose limits:** The effective doses to the individuals shall not exceed the limits recommended by the commission.

Radiation Dose Limits

The ICRP introduced the concept of permissible dose, defined as a dose of ionizing radiation that in the light of present knowledge is not expected to cause appreciable bodily injury to a person at any time during his lifetime. The recommended dose limits are given in the Table 11.4.

The occupational dose limits exclude exposures from medical procedures and natural background. The actual exposure to radiology department staff is relatively low. The effective dose is about less than 1 μSv per year for X-ray technologist. In the case of pregnant workers, the fetus is considered as a member of public. Pregnant radiation workers are monitored by a dosimeter worn on the abdomen under the lead apron. A measured dose of 2 mSv to the surface of the abdomen is normally considered equivalent to 1 mSv to the fetus. The dose limits for members of the public are generally 10 times lower than those for occupational exposure.

Table 11.4: Recommended dose limits

Applications	Occupational worker (mSv/year)	Public (mSv/year)
Effective dose	20*	1*
Eye lens	150 (modified as 20, in 2011)	15
Skin	500	50
Hands and feet	500	50
Fetus, mean dose	2 after diagnosis†	–

Source: International Commission on Radiological Protection (ICRP)-60, 1990.
*Averaged over a defined period of 5 consecutive years (1 mSv = 100 mrem), as per AERB it should not exceed 30 mSv in a single year; †After diagnosis, the limit to the surface of the women's abdomen is 1 mSv for the remainder of the pregnancy.

Dose Limits to Pregnant Women

In the case of women worker of reproductive age, once pregnancy has been established, the conceptus shall be protected by applying a supplementary equivalent dose limit to the surface of the women's abdomen (lower trunk) of 2 mSv for the remainder of the pregnancy. Internal exposures shall be controlled by limiting intakes of radionuclides to about 1/20 of annual limit of intake (ALI). The employment shall be of such a type that it does not carry a probability of high accidental doses and intakes.

Dose Limits for Trainees

Trainees age limit should be above 16 years and the annual dose limit is 6 mSv. No person under the age of 18 years shall be allowed to work in controlled areas unless supervised and only for the purpose of training.

Methods of Radiation Control

The three principal methods by which radiation exposures to persons can be minimized are:
- Time
- Distance
- Shielding.

Time

The total dose received by a radiation worker is directly proportional to the total time spent in handling the radiation source. Lesser the time spent near the radiation source, lesser will be the radiation dose. As the time spent in the radiation field increases, the radiation dose received also increases. Hence, minimize the time spent in any radiation area. Techniques to minimize time in a radiation field should be recognized or practiced.

All radiation sources do not produce constant exposure rates. Diagnostic X-ray machines typically produce high exposure rates over brief time intervals. For example, chest X-ray produces a skin entrance exposure of 20 mR in less than 1/20 of a second, equivalent to 1,440 R per hour. Hence, radiation exposure can be minimized by not energizing the X-ray tube, when personnel are nearer to the machine.

Nuclear medicine procedure produces lower exposure rate for extended periods of time. The exposure rate at 1 m from a patient injected with 20 mCi of Tc-99m, for bone scan is 1 mR per hour. It reduces to 0.5 mR per hour after 2 hours due to decay and urinary excretion. Hence, both the knowledge of exposure rate and how it changes with time are the important elements in reducing personnel exposures.

The time spent near the radiation source can be minimized by understanding the task to be performed and the suitable equipment to complete the task in short interval with safety. Hence, one has to plan the radiation procedure, practice the procedure without radiation and share the essential duties to reduce radiation exposure. For example, fluoroscopy screening time should be kept short by the use of last frame hold facility, in addition to the use of foot switch.

Worked example 1

A radiation worker is performing a procedure involving radiation and the equipment is 'ON' for 5 minutes for each examination. The radiation level at the location of the radiation worker is 60 mR/h. How many such procedures he/she can carry out per week?

The annual equivalent dose limit prescribed for the radiation worker is (occupational worker) 20 mSv

= 2,000 mrem = 2,000 mR

The permitted weekly dose

= 2,000 mR/50 week

= 40 mR

Exposure rate at the location of the worker

= 60 mR/h

= (60/60) mR/min = 1 mR/min

The exposure in each procedure

= (1 mR/min) × 5 min

= 5 mR

Hence, the number of procedures the worker can associate within 1 week

= 40 mR/5 mR

= 8

Distance

Radiation intensity (exposure rate) from a point source decreases with distance, due to divergence of the beam. It is governed by the inverse square law, which states that the exposure rate from a point source of radiation is inversely proportional to the square of the distance. If the exposure rate is X_1 at distance d_1, then the exposure rate X_2 at another distance d_2 is given by:

$$X_2 = X_1 \, (d_1/d_2)^2 \quad\quad\quad ----- (3)$$

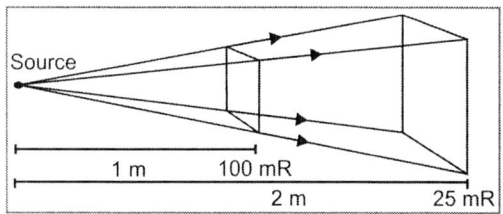

FIGURE 11.2: Inverse square law

Let, 100 mR is the radiation exposure at 1 m for a point source (Fig. 11.2). The radiation exposure at 2 m is found to be 25 mR, by inverse square law. Hence, if distance is doubled, the radiation is reduced by a factor of 4. Keeping larger distance always reduce radiation exposure. Larger the distance, lesser will be the radiation dose. This relationship is valid for point sources only, whose dimensions are very small compared to distance under consideration. Thus, the relationship is not valid near (< 1 m) a patient injected with radioisotopes, since the exposure rate decreases less rapidly than inverse square law.

In diagnostic radiology at 1 m from a patient, the scattered radiation is about 0.1–0.15% of the intensity of the primary beam. Hence, all personnel should stand as far away as possible during X-ray procedures. Personnel should stand at least 2 m from the X-ray tube and the patient and behind the shielded barrier or out of the room, whenever possible.

Imaging rooms should be designed to maximize the distance between the source and control console. Unshielded radiation sources never be manipulated by hand. Tongs or other handling devices should be used to increase the distance between source and hand.

Worked example 2

The exposure rate from a radiation source is 5 R/min at 50 cm. What would be the exposure rates at 40 cm and 60 cm respectively?

$X_1 = 5$ R/min
$D_1 = 50$ cm
$D_2 = 40$ cm
$X_2 = ?$
$X_2 = [X_1 \times (D_1)^2]/(D_2)^2$
$ = [5 \text{ R/min} \times (50 \text{ cm})^2]/(40 \text{ cm})^2$
$ = 7.81$ R/min
$X_1 = 5$ R/min
$D_1 = 50$ cm
$D_2 = 60$ cm
$X_2 = ?$
$X_2 = [5 \text{ R/min} \times (50 \text{ cm})^2]/(60 \text{ cm})^2$
$ = 3.47$ R/min

Answer: The radiation exposure rate at 40 cm and 60 cm are 7.81 R/m and 3.47 R/m respectively.

Shielding

When maximum distance and minimum time do not ensure an acceptably low radiation dose, adequate shielding must be provided, so that radiation beam will be sufficiently attenuated. The material that attenuates the radiation exponentially is called shield and the shield will reduce exposure to patients, staff and the public. If I_0 is the intensity of radiation at a point without shield and I is the intensity with a shield of thickness t, then, $I = I_0 e^{-\mu t}$, where μ is the linear attenuation coefficient of the shielding material. The thickness of the shielding material that reduces the intensity to half is called half-value layer (HVL) = $0.693/\mu$. Hence, larger the shielding thickness, lesser the radiation exposure.

X- and γ-rays undergo exponential attenuation in the shielding material. This means that even a large shielding material will not attenuate the radiation to zero intensity. However, optimal shielding thickness is required to bring down the radiation level below the permissible limit. Brick and concrete are used as shielding material for construction of X-ray room barriers. On the other hand, lead is used as protective material in lead apron, thyroid shield, viewing window and gonad shield. These protective shields usually reduce the radiation levels to 90–99%.

REGULATORY REQUIREMENTS

Atomic Energy Regulatory Board

The Atomic Energy Regulatory Board (AERB) is an apex body that regulates the use of ionizing radiation in India. The Chairman, AERB is the competent authority. The mission of AERB is to ensure that the use of ionizing radiation and nuclear energy does not cause undue risk to health and environment. It will develop and publicize specific codes and guides for radiation safety. The AERB implement the safety provisions by the Atomic Energy (Radiation Protection) Rules-2004, under the Atomic Energy Act (1962), which provides necessary regulatory infrastructure for effective implementation of radiation protection program in India.

Regulatory Requirements

Design Certification

All radiation equipments shall meet the design safety specifications stipulated in the AERB safety code. The manufacturer/vendor shall obtain design certification from the competent authority prior to manufacturing the radiation generating equipment.

Type Approval/ No Objection Certificate

Prior to marketing the radiation equipment the manufacturer shall obtain a Type approval certificate from the competent authority for indigenously made equipment. For equipment of foreign make, the importing/vending agency shall obtain a No Objection Certificate (NOC) from the competent authority, prior to marketing the equipment. Only Type approved and NOC validated equipment shall be marketed in the country.

Layout Approval and Registration

Once the radiation equipment is installed, it should be registered with AERB along with lay out approval. For registration and license, quality assurance tests and nomination of Radiological Safety Officer (RSO) is mandatory. This can be done with online through the AERB website (www.aerb.gov.in) called e-LORA (electronic licensing of radiation applications). e-LORA advocates institution registration, radiation professional (RP) registration. Only registered radiation professional can be nominated as RSO.

Inspection of Radiation Installations

The radiation installations shall be made available by the employer/owner for inspection at all reasonable times to the competent authority or its representative to ensure compliance with the safety code.

Decommissioning of Radiation Installations

Decommissioning of radiation equipment shall be registered with the competent authority immediately by the employer/owner of the equipment.

Certification of RSO

Any person accepting assignment to discharge the duties and functions of RSO for radiation equipment or installations shall do so only after obtaining certification from the competent authority for the purpose.

Certification of Service Engineers

Only persons holding valid certificate from the competent authority shall undertake servicing of radiation equipment.

Occupancy in the Room

Only persons whose presence is necessary should be in the radiation room during the exposure. Overcrowding should be avoided. All such persons must be protected with lead aprons or shields. The radiation room shall be kept closed during the radiation exposure.

Log Book

All X-ray equipment must have a separate log book, which provides information about the equipment manufacturer, model, serial number, date of purchase, cost, nature of fault, repair details, service personnel report, down time, etc.

Records

Records of all radiological examinations should be maintained. Reports and radiographs should be given to the patient for future reference.

Placard

A warning placard as shown in the Figures 11.3A and B must be posted outside the room entrance or door.

Atomic Energy (Radiation Protection) Rules-2004

Responsibility of the Employer (Rule 20)

1. Every employer shall:
   a. Ensure that provisions of these rules are implemented by the licensee, RSO and other worker(s).
   b. Provide facilities and equipment to the licensee, RSO and other worker to carry out their functions effectively.
   c. Obtain dose records and health surveillance report of the workers from their former employer.
   d. Provide dose records and health surveillance reports of the worker to the new employer.
   e. Furnish to each worker dose records and health surveillance reports of the worker annually.
   f. Inform the competent authority if the licensee or the RSO or any worker leaves the employment.
   g. Arrange for health surveillance of workers.
2. The employer shall be the custodian of radiation sources in his/her possession and shall ensure physical security of the sources at all times.

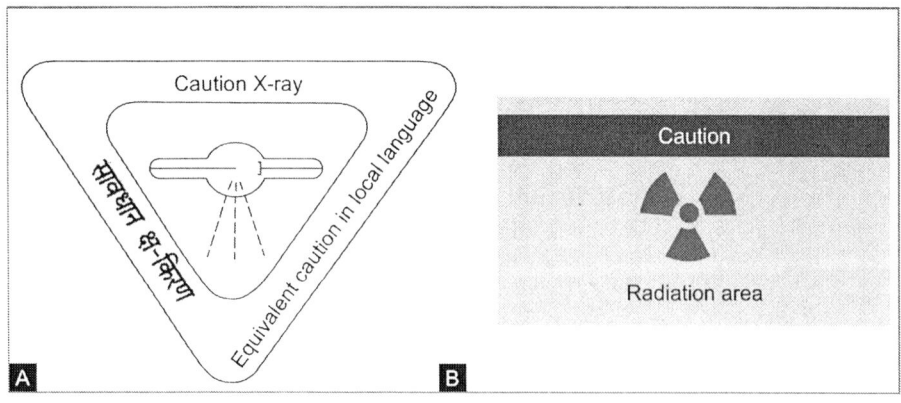

FIGURES 11.3A and B: Radiation warning placard. A. X-radiation warning sign; B. Gamma (γ)- rays warning sign.

3. The employer shall inform the competent authority within 24 hours, of any accident involving a source or loss of source of which he/she is the custodian.

Responsibility of the Licensee (Rule 21)

1. The licensee is responsible for the implementation of terms and conditions of the licensee.
2. The licensee shall comply with the surveillance procedures, safety codes and safety standards, specified by the competent authority.
3. Licensee shall establish written procedures and plans for controlling, monitoring and assessment of exposure for ensuring adequate protection of workers, members of the public and the environment and patients.
4. The licensee shall comply with the provision of rules for safe disposal of radioactive waste.
5. The licensee shall:
   a. Maintain records of workers.
   b. Arrange for preventive and remedial maintenance of radiation protection equipment and monitoring instruments.
   c. Investigate excessive radiation exposure and maintain records of such investigations.
   d. Inform competent authority about the occurrence, investigation and follow-up actions including steps to prevent future occurrences.
   e. Carry out physical verification of radioactive material periodically and maintain inventory.
   f. Inform appropriate law enforcement agency in the locality of any loss of source.
   g. Inform the employer and the competent authority of any loss of source.
   h. Investigate and inform the competent authority of any accident involving source and maintain record of investigations.
   i. Verify the performance of radiation monitoring systems, safety interlocks, protective devices and any other safety systems in the radiation installation.
   j. Prepare emergency plans in consultation with RSO.
   k. Conduct quality assurance tests of structures, systems, components and sources and related equipment.
   l. Advise the employer about the modifications in working condition of a pregnant worker.
   m. Inform the competent authority if the RSO or a worker leaves the employment.
   n. Inform the competent authority when he/she leaves the employment.
6. The licensee shall ensure that the workers are familiarized with contents of the relevant surveillance procedures, safety standards, safety codes, safety aides and safety manuals issued by the competent authority and emergency response plans.

Responsibility of Radiological Safety Officer (Rule 22)

1. Radiological Safety Officer (RSO) shall provide advice and assistance to the employer and licensee on radiation safety.
2. RSO shall:
   a. Carry out measurements and analysis on radiation and radioactivity levels in the controlled area, supervised area and maintain records of the same.
   b. Investigate any situation that could lead to potential exposures.
   c. Advice the employer to ensure regulatory constraints and the terms and

conditions of the license, safe storage and movement of radioactive material within the radiation installation to initiate suitable remedial measures in any situation that leads to potential exposures, and regular measurements and analysis of radiation and radioactivity levels in and around the installation.
   d. Report all hazardous situations with details and remedial actions taken to the employer and licensee for reporting to the competent authority.
   e. Conduct quality assurance tests on structures, systems, components and sources.
   f. Ensure periodic calibration of monitoring instruments.
3. RSO should assist the employer in:
   a. Instructing the workers about hazards of radiation and safety and good work practices.
   b. Safe disposal of radioactive wastes.
   c. Developing emergency response plans to deal with accidents and maintaining emergency preparedness.
4. RSO should advise the licensee on:
   a. Modifications in working condition of a pregnant worker.
   b. The safety and security of radioactive sources.
5. He/She should furnish to the licensee and the competent authority the periodic reports on safety status of the radiation installation.
6. He/She should inform the competent authority, whenever he/she leaves the employment.

Responsibility of the Worker (Rule 23)

1. Every worker shall observe safety requirements and follow safety procedures and instructions. He/She should not do any work that is harmful to him/her, co-workers, installation and public.
2. Worker should inform the employer about his/her previous occupations. He/She should use protective equipment, radiation monitors and personnel monitoring devices. He/She should inform the licensee and the RSO about accident or any potentially hazardous situation.
3. Female worker, once become pregnant should inform the same to the licensee and RSO.

Health Surveillance of Workers (Rule 25)

1. Every employer shall provide the services of a physician with appropriate qualifications to undertake occupational health surveillance of classified workers.
2. Every worker initially on employment and classified worker, thereafter at least once in 3 years as long as the individual is employed, shall be subjected to the following:
   a. General medical examination.
   b. Health surveillance to decide on the fitness of each worker for the intended task.
3. The health surveillance shall include:
   a. Special tests or medical examinations as specified by order by the competent authority for workers, who have received dose in excess of regulatory constraints.
   b. Counseling of pregnant workers.

PERSONNEL MONITORING

Thermoluminescent Dosimeter

The principle of radiation safety emphasize that each radiation worker's personnel dose should be within the dose limits. Hence, it is essential that the doses of the individual worker should be measured accurately. The system used for such measurements are called personnel monitoring systems. In India, Bhabha Atomic Research Centre (BARC) is conducting the personnel monitoring service for all the workers in the country. Nearly about 65,000 workers are monitored and their doses are maintained. Film badges were used initially to measure doses and it is replaced by thermoluminescent dosimeters (TLDs). The special features of TLD are high sensitivity, reusability, cheaper cost, easy to make, independent of climatic conditions and indigenously made.

Thermoluminescent Dosimeter Principle

Thermoluminescent dosimeters are used as personnel monitoring devices. It is based on the phenomenon of thermoluminescence, the emission of light when certain materials are heated after radiation exposure. It is used to measure individual doses from X-, β- and γ-radiations. It gives very reliable results, since no fading is observed under extreme climatic conditions. The typical TLD badge consists of a plastic cassette in which a nickel-coated aluminum (Al) card is placed (Fig. 11.4). Within the card, there are three TLD disks, which record the radiation exposure. When the TLD disk is exposed to radiation, the electrons in the crystal lattice are excited and move from the valence band to conduction band. Excited electrons, while returning to valence band are trapped just below the conduction band. The number of electrons in the trap is proportional to the radiation exposure, and thus, it stores the absorbed radiation energy in the crystal lattice. Once the card is exposed, to heat, the trapped electrons jump from traps to valence band by emitting light which is read by the TLD reader, which displays the dose in mSv. Usually, the service is offered for once in 3 months duration.

Thermoluminescent Dosimeter Card

Thermoluminescent dosimeter card consists of 3 $CaSO_4$:Dy-teflon disk of 0.8 mm thick and 13.2 mm diameter each and weighs about 280 mg. Teflon is used as binder that gives strength and does not absorb luminescence. The phosphor is mixed with Teflon in 1:3 ratio using liquid nitrogen and dried. They are sintered at 400°C in 1 hour before put it to use. The three disks are mechanically clipped over three symmetrical circular holes each of diameter 12 mm on a nickel plated aluminum plate (52.5 mm × 29.9 mm × 1 mm). An asymmetric V cut provided at one end of the card ensures a fixed

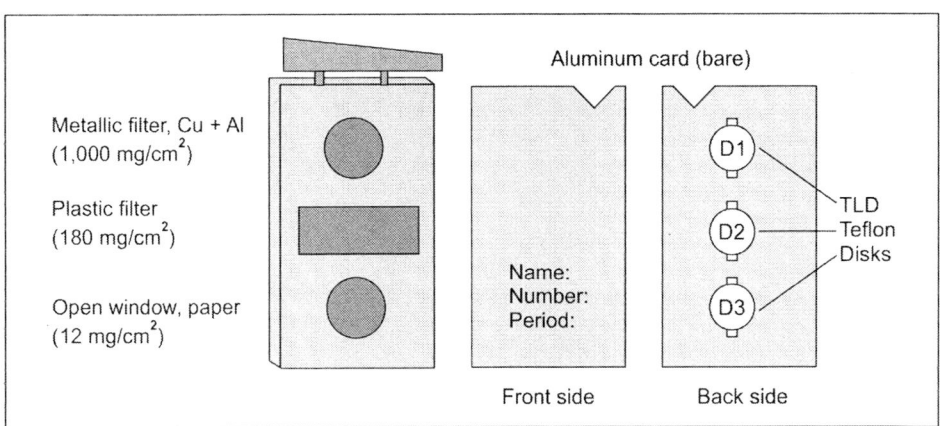

FIGURE 11.4: Thermoluminescent dosimeter (TLD) badge, aluminum cards and its holder with filters

orientation of card in the TLD cassette. The card is enclosed by a paper wrapper in which user's personnel data and period of use is written. The thickness of the wrapper (12 mg/cm^2) makes the measurements equivalent to 10 mm depth below the skin surface in the cassette. To protect the TLD disks from mishandling, the card along with its wrapper is sealed in a thin plastic (polythene) pouch. The pouch also protects the card from radioactive contamination, while working with open sources.

Thermoluminescent Dosimeter Cassette

Thermoluminescent dosimeter cassette is made of high impact plastic and has two parts, namely the base and the sliding parts with three circular regions. There are three filters in the cassette corresponding to each disk, namely Cu + Al filter, Perspex filter and open filter. Rectangular copper filter is fixed in the base, whereas the circular copper is fixed in the slider. The circular Al emblem is fixed both base and the slider. The above filter combination corrects energy dependence nature of the phosphor. A stainless steel crocodile clip is provided in the slider for easy attachment in user's clothing or to the wrist. TLD cassettes are available in the form of whole body (chest), wrist and forehead use.

When the TLD card is inserted properly in the cassette, the first disk (D1) is sandwiched between a pair of filter combination of 1 mm Al and 0.9 mm Cu (thick: 1000 mg/cm^2). The second disk (D2) is sandwiched between a pair of 1.5 mm thick plastic filters (180 mg/cm^2). The third disk (D3) is positioned under a circular open window. The metallic filter is meant for gamma radiation, and the Perspex is for β-radiation.

After radiation exposure, the dose measurements are made by using a TLD reader (Fig. 11.5). The reader has heater, photo multiplier tube (PMT), amplifier and a recorder. The reader is usually semi-automatic N$_2$ gas heating reader and the entire operation of the reader is controlled by a computer. The TLD cards are loaded in the magazine or planchet, where it is heated for a reproducible heating cycle. While heating, the electron returns to their ground state with emission of light. This emitted light is measured by the PMT, which converts light into an electrical current (signal). The PMT signal is then amplified and measured by a recorder. The reader is calibrated in terms of mR or mSv, so that one can get direct dose estimation. It can read about 50 cards in about 90 minutes. The measured doses are recorded as hard copy and stored.

Nowadays windows based computer-controlled TLD readers are available. They are

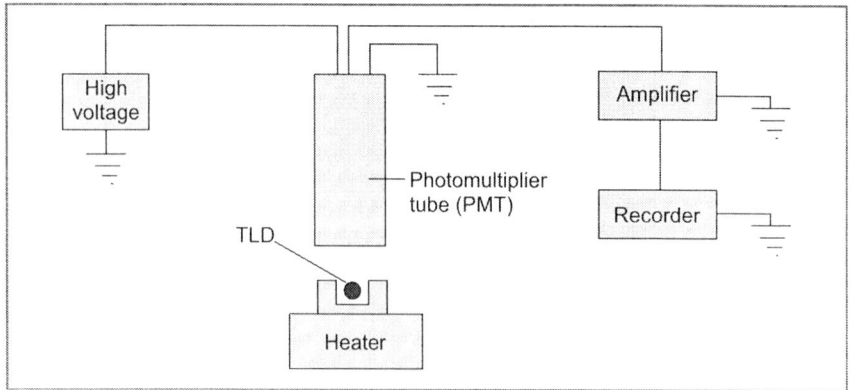

FIGURE 11.5: Thermoluminescent dosimeter reader

capable of analyzing TLD chips, ribbons, powder, disks, pellets, rods and micro cubes. They display digital glow curve and temperature profile. They can handle one or more planchets at a time either with manual drawer or computer-controlled drawer function. Programmable annealing oven is also available along with the system. In India, $CaSO_4$:Dy-teflon disks are used in country wide personnel monitoring.

The disks are reusable after proper annealing up to 300 times. The annealing process release the residual energy stored from the earlier exposure. A typical annealing cycle consists of 400°C for 1 h followed by 300°C for 3 h. This badge can cover a wide range of dose from 0.1 mSv to 100 Sv with a accuracy of ± 10%. TLD badges do not provide a permanent record. LiF can also be used as TLD phosphor, which has wide dose response over 10 μSv to 1,000 Sv. Its effective atomic number is close to that of tissue with an accuracy of ± 2%.

TLD badges are normally worn at the chest level, so that it is expected to receive the maximum radiation exposure to the whole body. Some of the radiation workers used to wear the badge at the waist level, which is not correct. During fluoroscopy, it is preferable for the radiologist to wear at the collar level to measure the dose to the thyroid and lens of the eye, since most of the body is shielded from the radiation exposure. Pregnant radiation workers should wear a second badge at waist level (under the lead apron) to assess the fetal dose. Additional wrist badge is advised for procedures involving nuclear medicine, brachytherapy source handling and interventional radiology.

Guidelines for Using TLD Badge

1. Thermoluminescent dosimeter badges are to be used only by persons directly working in radiation. Administrators, dark room assistant, sweepers, etc. need not be provided with TLD badges.
2. TLD badge is used to measure the radiation dose. It does not protect the user from the radiation.
3. The name, personnel number, type of radiation (X or gamma), period of use, location on the body (chest or wrist), etc. should be written legibly in block letters on the front side of the badge.
4. A TLD badge once issued to a person should not be used by any other person.
5. Each institution must keep one TLD card loaded in a chest TLD holder as control, which is required for correct dose evaluation. It should be stored in a radiation free area, where there is no likelihood of any radiation exposure.
6. TLD badge should be worn compulsorily at the chest level. It represents the whole-body dose equivalent. If lead apron is used, TLD badge should be worn under the lead apron.
7. While leaving the premises of the institute, workers should deposit their badges in the place where control TLD is kept.
8. A badge without filter or damaged filter should not be used. It is replaced by a new holder.
9. Every radiation worker must ensure that the badge is not left in the radiation field or near hot plates, ovens, furnaces, burners, etc.
10. Every new radiation worker has to fill up the personnel data form and should submit it to the BARC accredited agency.
11. All the used or unused TLD badges should be returned after every service period (quarterly) in one lot so as to reach 10th of next month.
12. Contact for all correspondence regarding TLD badge service to the Officer In-charge, personnel dosimetry and dose record section, Radiological Physics and Advisory Division, Bhabha Atomic Research Centre, CT & CRS building, Anusakti Nagar, Mumbai- 400094.

BIBLIOGRAPHY

1. Atomic Energy (Radiation Protection) Rules. Mumbai: Atomic Energy Regulatory Board; 2004.
2. Atomic Energy Regulatory Board Safety Code. Medical diagnostic X-ray equipment and installations AERB/SC/MED-2 (Rev 1).Mumbai: Atomic Energy Regulatory Board; 2001.
3. Atomic Energy Regulatory Board Safety Code. Nuclear medicine facilities AERB/SC/MED-4 (Rev 1). Mumbai: Atomic energy regulatory board; 2001.
4. Ayappan P. Personnel Monitoring Service. Workshop on Radiological Safety for Medical Physicists. Trichirappalli: Bharathidasan University; 2010.
5. International Atomic Energy Agency. Radiation biology: A Handbook for Teachers And Students, Training course series 42. Vienna: International Atomic Energy Agency; 2010.
6. International Commission on Radiological Protection. Biological Effects After Prenatal Irradiation (Embryo and Fetus). ICRP publication 90, New York: Elsevier; 2003.
7. International Commission on Radiological Protection. Recommendations of the International Commission on Radiological Protection. ICRP report 103. New York: Elsevier; 2007. Ann ICRP 2007a:37:1-332.
8. Jerrold T Bushberg, Anthony Seibert J, Edwin M Leidholdt JR, et al. The Essential Physics of Medical Imaging, 3rd edition. Philadelphia, USA: Wolters Kluwer/Lippincott Williams & Wilkins; 2012.
9. Thayalan K. Textbook of Radiological Safety. New Delhi: Jaypee Brothers Medical Publishers (P) Ltd; 2010.
10. Thayalan K. The Physics of Radiology and Imaging. New Delhi: Jaypee Brothers Medical Publishers (P) Ltd; 2014.

Index

Page numbers followed by *f* refer to figure and *t* refer to table.

A
Anaphase 8
 bridge 51
Anoxic cells 70
Apoptosis 16
Apoptotic death 64
Aqueous electron 41*f*
Atomic
 energy
  regulatory board 175
  rules-2004 176
 number 21, 23*t*, 33, 33*t*
 structure 21, 21*f*
Autophagic death 64

B
Bergonié and Tribondeau law 48
Binding energies 23*t*
Biodosimetry 52
Bladder 107
Blister formation 102
Blood cell depression 103*f*
Brachytherapy 135, 137, 138*f*, 144, 156, 163*f*
 biological basis of 135
 head and neck 161*t*
 sources 136*t*
Bragg curves 35*f*
Brain 107
Bystander effect 63

C
Cancer 91*f*
 bone 93
 breast 93, 134
 cells 19, 19*t*
 cervix 147
 development 17, 18*f*
  steps in 18
 head and neck 133, 146
 induction 87
 lung 93
 radiation induced 87, 92
 skin 93
 staging 20
 thyroid 92
Carbohydrate 12, 13*f*
Carcinoma 17
 cervix 160*t*
Cell 4
 cycle 6, 7, 7*t*, 48
  checkpoints 8
  stage 76
 death 9, 54
  mechanism of 63
 function 6
 lines, survival curve of 137*f*
 radiosensitivity of 70*f*
 structure 4, 4*f*
 survival
  curves 54
  modification of 68
Chromosomal aberration analysis 52
Chromosome 6
 damage 50
  radiation induced 51*f*
 replicate 7*f*
Circulatory system 3
Compton scattering 27, 29*f*, 30, 31*f*
Consciousness 111
Conventional fractionation 130
Coulomb force 15
Crosslink repair 62
Cube root rule 149
Cumulative radiation effect 151
Cutaneous system 115
Cysteine 78*f*
Cytoplasm 5

D
Deoxyribonucleic acid (DNA) 12, 13*f*, 14*f*, 40*f*, 54
 damage induced nuclear foci 45
 double strand break repair 62
 molecule 13*f*
 stability gene 17
 strand break measurement 45
 functions of 14
Diarrhea 111

Digestive system 3
Dose response
    curves 127
    models 88
Dry desquamation 102

E

Elastic collision 35, 36
Electromagnetic
    radiation 24, 26t
    spectrum 25, 25f
    wave 24f
Electron 21f
    interaction 35
    volt 23
Endocrine system 3
Endoplasmic reticulum 5
Energy dependence 28
Epilation 102
Erythema 102
    stage 101
    late 101
Esophagus 104
Excitation 23, 41
External beam
    radiation therapy 160t
    radiotherapy, biological basis of 120
Eye lens 106

F

Fetal period 118
Fluorescence 26
Free radical reaction 41

G

Gastrointestinal
    syndrome 112
    system 116
    tract 104
Gene mutation hypothesis 17
Genetic diseases 94
Genome 14
Giant cell formation 49
Golgi complex 5
Growth
    disorder factors 16
    process 16
Gynecological brachytherapy 157t, 162

H

Headache 111
Heart 108
Heating methods 82
Hematopoietic syndrome 111
Hematopoietic system 102, 115
Human
    anatomy 1
    biology 1
    body composition 9
    peripheral blood lymphocytes 53f
Hydrogen
    bond 16
    radical 41
Hydroxyl radical 41
Hyperfractionation 131
Hyperplasia 16
Hyperthermia 80, 82
Hypertrophy 16
Hypofractionation 133
Hypoxia
    acute 71
    chronic 70

I

Integumentary system 3
Intestinal
    pathology 113
    villi 105f
Intestine, small and large 104
Inverse square law 26, 174f
Ionization 23, 40
Ions 9
Isotopes 22

K

Kidney 107

L

Lethal type chromosome aberrations 50
Leukemia 92, 92f
Linear
    attenuation coefficient 28
    energy transfer 73
        values 74t
    quadratic
        model 153
        theory 59
Lipids 11
Liver 107
Lung 107
Lymphatic system 3
Lymphomas 17
Lysosomes 5

M

Macromolecule
    irradiation of 43
    radiation damage of 43f

Malignant
    cell 19f
    tumors 17
Mammalian cells 55f, 57, 59f, 81f
    cell cycle of 7f
    survival response of 78f
Mass
    attenuation coefficient 28
    density 33
    energy equivalence 25
    number 21
Meiosis 8
Metaplasia 16
Microtubules 6
Mismatch repair 62
Mitochondria 5
Mitosis 7, 7f
Mitotic
    cell death 64
    inhibition 47
    mutations 46
Molecules 10
Multichromosomes 17
Multifactorial diseases 95
Multitarget theory 57
Muscular system 3
Mutation 17, 46

N

Necrosis 102, 150f
Neoplasia 16
Neoplasm 16
Nervous system 3
Neurovascular syndrome 113
Neutron 21f, 23
    capture 37
    interaction 36
    properties of 24
Non-ionizing radiation 25
Non-lethal
    damage 67
    type aberrations 51
Normoxic cells 70
Nuclear medicine 118
Nucleic acid 12
Nucleotide 13f
    excision repair 61
Nucleus 6

O

Oral mucosa 104
Organ 2
    systems 114t
Organogenesis 117

Ovary 106
Oxygen
    concentration 69
    effect 68

P

Partial body irradiation 103
Phosphate molecules 13f
Photoelectric absorption 27, 30, 31f
Photon energy 34
    function of 31f
Photonuclear disintegration 27
Plasma membrane 4
Postsurgery radiotherapy 133
Pregnancy, termination of 118
Protein 10, 21f
    primary structure of 11f
Proton interaction 36
Proto-oncogene 17
Pulsed-field gel electrophoresis 45

Q

Quantum number 22

R

Radiation 50, 54
    accidents, management of 114
    biological effects of 84, 84f
    carcinogenesis 86
    control, methods of 173
    damage
        classification of 65
        repair of 61
    dose
        acute 85
        chronic 85
        limits 172
    effects 47, 84, 99, 101, 114
    exposure, acute 110f
    installations 175
    interaction 27, 38f
    ionizing 25
    normal tissue reactions of 99
    physics 21
    protection 168
        principles of 172
    quality 87
    safety 168
    sensitizers 79
    sources of 170
    syndrome 108, 109f, 111t
    treatment 164
    units 26
    weighting factors 169t

Radioprotectors, structure of 78f
Radiotherapy 71, 79, 119, 133
    evaluation of 120
Reoxygenation 125, 140
Reproductive
    organs 105
    system 3
Respiratory system 3
Ribonucleic acid (RNA) 12, 13f
Ribosomes 5

S

Sarcoma 17
Single cell electrophoresis 45
Single hit-multi target theory 57, 58
Single hit-single target theory 58
Skeletal system 3
Skin 101
    carcinoma cure 150f
    dry desquamation of 150
    erythema 150f
        dose 101
    injury 102t
    layers of 101f
    moist desquamation of 150
Spinal cord 108
Stem cells 105f
Stomach 104
Survival curve 55, 59f, 81f
    relevance of 60
    shape 69f

Synchronous cell
    cultures 49
    population, formation of 50f
Synthesis period 7f

T

Thermoluminescent dosimeter 178
    card 179
    cassette 180
    principle 179
    reader 180f
Tissue 2
    growth 2
    weighting factors 169t
TNM classification 20
Total body irradiation 102
Tumor
    benign 17
    biology 16
    cells 124f
    oxygen effect in 70
    suppressor gene 17
Typical hyperthermia system 82f

U

Ulceration 102
Urinary system 3

V

Van der Waals force 16
Vomiting 111

Textbook for OPHTHALMIC ASSISTANTS

As per the Latest Guidelines and Syllabus

PK Mukherjee MBBS MS
Practicing Consultant and Counselor
Department of Ophthalmology

Formerly
Professor and HOD in the Upgraded Department of Ophthalmology
Regional Institute of Ophthalmology at Pt Jawahar Lal Nehru Memorial Medical College
Raipur, Chhattisgarh, India

JAYPEE BROTHERS MEDICAL PUBLISHERS
The Health Sciences Publisher
New Delhi | London

 **Jaypee Brothers Medical Publishers (P) Ltd**

Headquarters
Jaypee Brothers Medical Publishers (P) Ltd
EMCA House, 23/23-B
Ansari Road, Daryaganj
New Delhi 110002, India
Landline: +91-11-23272143, +91-11-23272703
+91-11-23282021, +91-11-23245672
Email: jaypee@jaypeebrothers.com

Corporate Office
Jaypee Brothers Medical Publishers (P) Ltd
4838/24, Ansari Road, Daryaganj
New Delhi 110 002, India
Phone: +91-11-43574357
Fax: +91-11-43574314
Email: jaypee@jaypeebrothers.com

Overseas Office
J.P. Medical Ltd
83 Victoria Street, London
SW1H 0HW (UK)
Phone: +44 20 3170 8910
Fax: +44 (0)20 3008 6180
Email: info@jpmedpub.com

Website: www.jaypeebrothers.com
Website: www.jaypeedigital.com

© 2024, Jaypee Brothers Medical Publishers

The views and opinions expressed in this book are solely those of the original contributor(s)/author(s) and do not necessarily represent those of editor(s) and publisher of the book.

All rights reserved. No part of this publication may be reproduced, stored or transmitted in any form or by any means, electronic, mechanical, photocopying, recording or otherwise, without the prior permission in writing of the publishers.

All brand names and product names used in this book are trade names, service marks, trademarks or registered trademarks of their respective owners. The publisher is not associated with any product or vendor mentioned in this book.

Medical knowledge and practice change constantly. This book is designed to provide accurate, authoritative information about the subject matter in question. However, readers are advised to check the most current information available on procedures included and check information from the manufacturer of each product to be administered, to verify the recommended dose, formula, method and duration of administration, adverse effects and contraindications. It is the responsibility of the practitioner to take all appropriate safety precautions. Neither the publisher nor the author(s)/editor(s) assume any liability for any injury and/or damage to persons or property arising from or related to use of material in this book.

This book is sold on the understanding that the publisher is not engaged in providing professional medical services. If such advice or services are required, the services of a competent medical professional should be sought.

Every effort has been made where necessary to contact holders of copyright to obtain permission to reproduce copyright material. If any have been inadvertently overlooked, the publisher will be pleased to make the necessary arrangements at the first opportunity.

Inquiries for bulk sales may be solicited at: jaypee@jaypeebrothers.com

Textbook for Ophthalmic Assistants

First Edition: **2024**

ISBN: 978-93-5696-545-4

Printed at: Sterling Graphics Pvt. Ltd.